PLANT TECHNOLOGY
OF FIRST PEOPLES IN BRITISH COLUMBIA

ROYAL BC MUSEUM HANDBOOK

PLANT
TECHNOLOGY
OF FIRST PEOPLES IN BRITISH COLUMBIA

NANCY J. TURNER

ROYAL **BC** MUSEUM
Victoria, Canada

Library and Archives Canada Cataloguing in Publication Data
Turner, Nancy J., 1947–
    Plant technology of First Peoples in British Columbia
    (Royal BC Museum handbook)

    Previously issued 1998 by UBC Press for the Royal British
        Columbia Museum.
    Includes bibliographical references: p.
    ISBN 978-0-7726-5847-0

        1. Indians of North America – Ethnobotany – British Columbia –
Handbooks, manuals, etc. 2. Indians of North America – Material
culture – British Columbia – Handbooks, manuals, etc. 3. Plants,
Useful – British Columbia – Handbooks, manuals, etc. 4. Botany,
Economic – British Columbia – Handbooks, manuals, etc. 5. Plants –
British Columbia – Handbooks, manuals, etc. I. Royal BC Museum.

E78.B9 T87 2007   581.6'36089970711   C2007-960196-0

CONTENTS

PREFACE TO THE FIRST EDITION

This is the third handbook of the British Columbia Provincial Museum dealing with the uses of plants by the First Peoples of British Columbia. The previous handbooks covered the food plants – the first on coastal peoples and the second on interior peoples. This volume concerns plants used in aboriginal technology. I have combined the uses of plant materials by coastal and interior peoples, because there is considerable overlap between the coast and interior in the species of plants used for technological purposes.

As a child I used to watch with fascination Kwakwaka'wakw Chief Mungo Martin at Thunderbird Park in Victoria hewing massive animal forms from Western Red-cedar logs with nothing more than a small hand-adze and a chisel. From that time on, my respect and appreciation for the knowledge and technological skills of First Peoples has grown. Examining the delicate baskets with their intricate designs, the perfectly symmetrical canoes, the tools and fishing equipment, the kerfed boxes, the giant feast dishes, and any number of other items to be seen in museums and private collections, you can't help but be impressed with the workmanship and industry of the men and women who made them. And you get an even greater appreciation if you try to make these items yourself.

My first effort at imitating aboriginal technology was to make a small Cattail mat, under the kind guidance and direction of Christopher Paul of the Tsartlip Reserve at Brentwood Bay. By the time I finished gathering and preparing the materials for this modest effort and constructing it according to traditional techniques, my admiration for aboriginal artisans had increased tenfold. Later, at Massett on Haida Gwaii, Florence Davidson taught me how to make twined cedar-bark baskets, and guided me through the long and arduous procedures of collecting, splitting and drying the bark, then weaving it in the proper way. My basket took many days to complete and looked like the work of a little child compared to her beautiful cedar-bark hats and baskets, but I was quite proud of my efforts, and once again my appreciation of aboriginal artisans, both past and present, took a giant step forward.

Many people are inclined to believe that aboriginal technological skills are a thing of the past. It is true that the number of items being made has decreased considerably within the last century, and that many of today's aboriginal craftspeople are of the oldest generation. Some skills, such as the old-time techniques of dyeing with natural mineral and vegetable dyes, have almost completely vanished. Nevertheless, young people in First Nations communities are becoming increasingly aware of the rich cultural heritage of past generations, and the future of many technological arts looks bright indeed. By imitating the works of their ancestors, as seen now mainly in museums, modern aboriginal artists, particularly carvers, have been able to recapture and even improve upon the skills and techniques of the past. First Nations artists still make dishes, spoons, masks, kerfed bent-wood boxes, and beautifully imbricated baskets of spruce root, split cedar root, sedges and other materials, which they sell privately or through retail outlets (such the Royal Museum Shop).

I have been privileged to learn about plants used in aboriginal technologies from many knowledgeable and respected members of First Nations communities in the province. They generously contributed their time and expertise to my studies, and I am forever indebted to them. I have already mentioned Christopher Paul of Tsartlip and Florence Davidson of Massett. Others are: Emma Matthews and William Matthews (Chief Weah) of Massett; George Young, Maude Moody, Agnes Moody, Solomon and Emma Wilson, and Becky Pearson of Skidegate; Gertrude Kelly of Vancouver; Margaret Siwallace, David Moody and Felicity Walkus of Bella Coola; Cecelia August of Sechelt; Andy Natrall Sr, Louis Miranda and Dominic Charlie of North Vancouver; Elizabeth and Richard Harry of the East Saanich Reserve; Lucy Brown, Daisy Roberts, Agnes Alfred and Agnes Cranmer of Alert Bay; Bob Wilson, and Mr and Mrs Tom Johnson of Fort Rupert; George Ignace, Mike Tom and his wife, "Mrs Mike", and Alice Paul and her son, Larry Paul, of Hesquiaht; Luke Swan of Manhousaht; Sam Mitchell of Xaxl'ip (Fountain); Martina LaRochelle of Lillooet; Charlie Mack Seymour and Baptiste Ritchie of Mount Currie; Annie York of Spuzzum; Louis Phillips of Lytton; Selina

Timoyakin, Larry Pierre, Lillie and Willie Armstrong, and Martin Louie of Penticton; Harry Robinson of Keremeos; Eliza Archie of Canim Lake; Aimee August of Chase; Frank Whitehead and Mary Paul of St Mary's Mission, Cranbrook; and Catherine Gravelle of Tobacco Plains. Without these people I could never have written this book. I sincerely hope they and their families enjoy it and view it with a feeling of pride.

I would also like to thank Randy Bouchard and Dorothy Kennedy of the British Columbia Indian Language Project, Victoria, for permission to include information on plants used by the Mainland Comox people which they recorded from Rose Mitchell, Bill Mitchell, John Mitchell and Jeannie Dominick of Squirrel Cove near Powell River, and for information on plants used by the Secwepemc, which they recorded from Aimee August and Isaac Willard of Chase. I am also grateful for the use of material from Kennedy and Bouchard's unpublished manuscripts on aboriginal fishing technologies. David Rozen of Victoria provided some information on the use of Black Hawthorn, which he obtained from Abel Joe of the Cowichan Reserve. He also compiled ethnobotanical information from the unpublished notes of Diamond Jenness, cited in the References. David Ellis of Vancouver, working with Luke Swan of Manhousaht, also contributed information used in this handbook. Dr Brent Galloway of the Coqualeetza Education and Training Centre in Sardis kindly provided information on Reed Canary Grass used for basketry by the Upper Sto:lo.

Many, many others, including all of the linguists I have had the pleasure of working with over the last few years, have contributed to this book in no small way. I would also like to acknowledge: Freeman King of Victoria for inspiration and encouragement; Dr Marcus Bell of the Biology Department at the University of Victoria, and Dr Roy Taylor, Director of the Botanical Garden at the University of British Columbia (UBC), for their support and advice during my university years when I carried out much of my research; Yorke Edwards, Director of the British Columbia Provincial Museum (BCPM), Dr Adam Szczawinski, former Curator of Botany, and Dr T.C. Brayshaw, Associate Curator of Botany at the BCPM, for advice and financial support; Dr R.T. Ogilvie, Curator of Botany at the BCPM, for his critical reading of the manuscript; my student assistant for two summers, Kathleen Cowen, for her enthusiastic dedication, especially during field trips; Harold Hosford of the BCPM and Dr Jack Maze of UBC's Botany Department for their editorial counsel; Peter Macnair and Alan Hoover of the Ethnology

Division and Drs Robert Levine and Barbara Efrat of the Linguistics Division of the BCPM, and Dr Andrea Laforet of the National Museum of Man in Ottawa, for their supplementary criticisms and suggestions; and my husband, Robert D. Turner, for his continuing moral support, and for contributing his photographic expertise to this project.

N.J.T.
1979

PREFACE TO THE SECOND EDITION

Since the first edition of this book (called *Plants in British Columbia Indian Technology*) was published in 1979, there have been many new developments and discoveries in the use of plant materials. There are active and growing centres of basket weaving, woodcarving and other arts using plant materials in different parts of British Columbia and surrounding areas. Many accomplished basket weavers and wood-carvers are teaching courses and giving workshops to an increasing number of enthusiastic learners at First Nations' education and cultural centres throughout the province. Notably, a number of aboriginal communities have embarked on special projects to carve large dugout canoes or, in the interior, to construct bark canoes. Some of these vessels have been in use for many years now.

Archaeologists have made many new discoveries about the uses of plant materials. Several wet sites along the coast have yielded a wealth of plant materials and of artifacts made from them or used in plant processing. The most famous of these is at Ozette, an ancient Makah village near Neah Bay on the Olympic Peninsula in Washington, where a number of houses and all of their contents and occupants were buried in mud slides from several earthquakes 400 to 600 years ago. These tragedies, like the deadly volcanic ash that covered the ancient city of Pompeii in Italy, ultimately turned out to have unexpected benefits. When it was discovered that the village site was being exposed and its houses and artifacts washed away by ocean waves, the Makah people, in collaboration with archaeologists (Dale Croes, among others) began an intensive excavation of Ozette. Their findings were truly astonishing. Because the site had been buried by water-saturated mud, the lack of oxygen prevented the usual bacterial and fungal decay of the plant materials that would normally occur in a wet temperate environment. Plant materials uncovered from Ozette include: woven cedar-bark matting and cedar-bark pouches for whaling harpoons, cedar-withe baskets, large quantities of cordage, looms and spindle whorls, hats, parts of garments, cedar planks, posts and canoes, Yew-wood implements, nettle-fibre fishnets, and many other articles of wood and fibre, made with amazing skill and artistry. All these artifacts were carefully salvaged, treated to prevent their decay, catalogued and housed in the Makah Museum at Neah Bay. For the Makah people, it was like a window into their past. They learned much about the antiquity of their material culture and, with the help of contemporary elders, were able to gain a better understanding of the materials and arts of their ancestors. A video called *Indian America – A Gift from the Past* (Media Resource Associates 1994) features the Ozette discovery and its outcomes. You can read about some of the research on Ozette and other wet sites on the coast in Bernick 1991 and 1998, Croes 1976, 1977, 1989 and 1995, and Kirk 1979 (all listed in References).

Dry-land archaeology, such as that undertaken by Brian Hayden and his research group at Keatley Creek near Lillooet, and by George Nicholas and his team at Kamloops, has also yielded some fascinating finds of past plant resource use (see: Mathewes 1980; Hayden 1991; Lepofsky in press, a, b; Schlick 1994; Nicholas 1997; Nicholas and Andrews 1997). In particular, archaeologists have discovered many rolls of birch bark, some obviously used in basketry and some for lining storage pits, dating back 2,000 years or more.

Some researchers have also undertaken a re-evaluation of tools found in archaeological sites, and have concluded that many, including awls, knives, root scrapers and bark peelers of stone, bone or antler, were likely used to process plant materials, not just in hunting or fishing activities. Working with fibrous plant materials – e.g., for making mats and baskets – has generally been the work of women, and as with the gathering of plants for food, has sometimes been under-recognized in archaeological and ethnographic descriptions. Today, the enhanced profile of plants and prehistoric plant use is both contributing to and being enhanced by an increasingly rich documentation of plant remains emerging from both wet and dry archaeological sites.

The archaeological evidence for basket making extends back thousands of years. Archaeological remains of basketry generally represent the same styles and techniques used by today's basket makers. The earliest findings of basketry at a wet site in British Columbia come from the Glenrose Cannery Site in the Fraser Delta region: 4,440 (± 80) years ago for wood basketry (withe or tree root) and 3,970 (± 90) years ago for bark basketry (Bernick 1991, 1998). Interestingly, one of the most important sources of basketry materials, Western Red-cedar, was not common as a species on the Northwest Coast until approximately 4,000 years ago, and did not occur until much later in some regions, such as the Skeena River (Hebda and Mathewes 1984).

Research on the social and symbolic aspects of plant-material use has also been ongoing. Andrea Laforet, who has studied basketry of B.C. First Peoples for many years, maintains that "the basketry style of a people encodes their identity" (Laforet 1990: 286). She has demonstrated significant, though sometimes subtle, distinctions in style and technique not only among different regions and cultural groups, but among individual weavers. She also notes that each basket maker had her own personal style and that "the making of any basket involved the basket maker in a creative tension between the fulfilment of the regional style and the expression of self" (Laforet 1990: 292).

The role of plant materials in trade has also become increasingly clear. Trading in woven and carved goods was well established in pre-European times. Croes (1977), for example, identified a northern-style woven hat at Ozette, dating to about 450 years ago. Nlaka'pamux elder Annie York described how some women, especially widows, wove baskets, bags and mats for a living, even long ago: "These women have no more husbands to go and gather fish. So that's the way they trade with the people from the interior. They make baskets – round baskets for whipping Soapberries, or berry baskets – and they trade them. And their mats, too. We use lots of those bulrushes, and the people from here [Spuzzum] like Silver Willow [Silverberry] bags. They bring them down here, too, and they trade them with baskets made out of split cedar."

First Peoples exchanged not only baskets and other materials, but also weaving techniques. Kennedy and Bouchard (1983) report that, in the late 1800s, Sliammon women travelling to Kamloops for Catholic prayer meetings learned the art of cedar-root basketry from Interior Salish women. At about the same time, Haida women learned false embroidery decoration techniques from their Tlingit neighbours (Blackman 1982). In their work with plant materials, people have had to adapt to new ideas and new markets. One example is the tea set made from coiled split cedar roots (see the photograph on the next page). Other modern examples include weavings around bottles and other containers, and trays, trunks and coffee tables.

Another new area of research, which in some ways has only just begun, is in First Peoples' traditional management practices for harvesting their plant resources sustainably. When you think about the quantities of plants harvested – more than 200 Tule stems to make one

Tea set made from coiled split cedar roots. A Salish weaver made this item to sell to tourists.
(RBCM collection)

small mat, for example, or a long strip of cedar bark to make a medium-sized basket – you can appreciate the tremendous volume of materials routinely harvested over past centuries. And yet, as Anderson (1993a, b, c) and others have demonstrated in their studies of basket materials in California, harvesting did not necessarily deplete the resources, as long as it was done with care. Techniques such as pruning and burning actually enhanced the quality and productivity of some plant materials. Basket makers in British Columbia say the same thing. Lena Jumbo of Ahousaht, a highly accomplished Northwest Coast basket maker, declares that Slough Sedge leaves are of better quality if cut from patches where they have always been harvested (see Craig and Smith 1997).

Stronger than ever, now, are my fascination for plant materials and my respect for the artists who work in wood and weave plant fibres into objects of beauty and utility. An important and enjoyable component of a fourth-year ethnobotany course that I teach at the University of Victoria is a manufacturing project in which students research and make an object from plant materials. This can be as basic as a string of Stinging Nettle fibre or as complex as a Cattail mat, canoe paddle or fire-making apparatus. Without exception, the students come away with a deepened appreciation of the skills and knowledge required to work with plant materials. In 1997, inspired by the teachings of mentors such as basket-makers Mary Thomas and Theresa Parker, we decided to collect some of the written descriptions of these projects into a manual, "Making it with Your Hands": Projects Using Indigenous Plant Materials from British Columbia (Turner 1998). This book and another, Plants for All Reasons (Turner 1992c), put together by class researchers in 1991, are available at the University of Victoria bookstore.

Many of my aboriginal friends and teachers who contributed to the first edition of this book have now passed away. Almost all of the elders acknowledged in the preface to the first edition are no longer living. I often think about them with great fondness, and I still marvel at all they knew and experienced in their lifetimes. I wrote about one inspiring weaver, Nellie Peters, in a short article entitled "'Just When the Wild Roses Bloom': The Legacy of a Lillooet Basket Weaver" (Turner 1992b). Florence Davidson's recollections of the spruce-root basketry of her mother are recorded in her biography, *During My Time* (Blackman 1982). Another book featuring Florence Davidson and other Haida weavers is a beautifully illustrated children's book called *The Weavers* by Jenny Nelson (1983). In fact, there are a number of children's books available on topics relating to plant use, such as *For Someone Special*, written and illustrated by Laura Boyd (1990), about Dakelh cradle-making. The Secwepemc Cultural Education Society (1986a, b, c) has also published a series of educational books on the technology of the Shuswap (Secwepemc) and related topics. Over the past several years, I have had the pleasure of working with the Secwepemc Cultural Education Society, together with Dr Marianne Ignace, Chief Ron Ignace, Dr George Nicholas and many Secwepemc elders and other researchers, on an ethnobotany project that will eventually yield a comprehensive book on Secwepemc Plant Knowledge, including traditional plant materials of the Secwepemc peoples.

The most important legacy of the basket weavers, carvers and workers in plant materials, however, is not in written words, but in their hands-on teachings, especially among their own families and communities. Primrose Adams and Virginia Hunter, who learned spruce-root basketry from their mother, Florence Davidson, have been teaching spruce-root weaving on Haida Gwaii for many years. Another Haida weaver, April Churchill, has also inspired many to learn basketry arts. Other communities are fortunate to have dedicated teachers who learned their arts from elders of the previous generations. William White, a Tsimshian weaver who gained much of his weaving expertise from his aunt, devotes much of his time to teaching others in Tsimshian and Haida communities about weaving with cedar bark and spruce roots, and creating Chilkat-style blankets (see the photograph on the next page). Another active and inspiring teacher, especially of birch-bark basketry, is Secwepemc elder Mary Thomas of the Neskonlith Band at Salmon Arm. She and her family have produced a video about the philosophy and practice of birch-bark basketry. Mandy Jimmie, a Nlaka'pamux mat weaver of the Nicola Valley, was inspired by earlier

Carrying on the tradition: Red-cedar-bark basket made by Tsimshian artist William White.

weavers like Mabel Joe. She and her colleagues are now reconstructing some of the early weaving patterns for Tule, Cattail and other species used in mat-making.

In the United States, a politically active and inspiring group of basket weavers founded the California Indian Basketweavers Association, with its own newsletter, workshops and annual meetings; in 1996, they released an educational video, *From the Roots: California Indian Basketweavers.* This organization has inspired others, such as the Northwest Basketweavers Association, established in 1996, with master basket weavers and interested members throughout Washington, Oregon, British Columbia and Alaska. Woodcarving is also thriving in various communities.

It is important to emphasize that the actual knowledge of weaving or carving techniques is a relatively small part of the entire knowledge needed: the times and locations for harvesting materials, ways to harvest without harming the plant populations, and the techniques for processing and storing the materials are just as critical to the production of the final products.

Although some knowledge and skills have faded, much has been retained and continues to grow and strengthen. Collaborations with museums (such as the Royal B.C. Museum and the Canadian Museum of Civilization) have proved extremely fruitful in terms of examining early collections of plant-based artifacts, and in highlighting the skills and knowledge of people in this area. A notable example is the recent exhibit at the Canadian Museum of Civilization that features Nlaka'pamux clothing and weaving technologies. Many examples of woven and carved objects, ranging from twined baskets to wooden masks and ocean-going canoes, are exhibited in museums, and many of these items are sold in museum gift shops throughout British Columbia.

In some ways plant materials have not had as high a profile as traditional food plants, but their harvesting and continued use, and all the important knowledge associated with their applications, are no less an integral part of First Peoples' cultural heritage and identity. The plants and the methods of harvesting and using them are embodied in complex and rich knowledge systems termed Traditional Ecological Knowledge, and this information is being incorporated into educational and cultural programs all over the province. Aboriginal leaders, educators and community members continue to work towards more control over their traditional lands and resources; and the specialty woods, fibres and other plant materials yielded by the land will continue to play a vital role in First Peoples' lives. The strategies, traditions, ceremonies and stories about cedar bark, Western Yew, birch, Indian Hemp, tree pitch, Douglas-fir boughs and all the other plant materials described in this book are integral to their ongoing sustainable use. The philosophies and world views represented by this use are as important now as at any time in the past.

A new area of interest in traditional plant materials is their increasing application as non-timber forest products to enhance the value of forests above and beyond the so-called "fibre values" of a few tree species. Diversifying our use of forest ecosystems is well recognized as a component of sustainable forestry practices. One Gitxsan leader talked about the "thousand-dollar birch tree," as an example of the different values placed on native plants by different sectors of society. In industrial forestry, birch is often viewed as a useless "weed" tree, competing with the more economically valuable conifers for space. If cut and used at all, it brings a minimal value for pulp. For Gitxsan and other First Peoples, though, a single birch tree can yield enough bark to make several baskets and enough wood to carve many spoons and other small implements that not only have cultural value, but can be marketed as "value-added" products.

Finally, native plants, whether for food or materials, provide exciting opportunities for gardening and landscaping. Using native plants in gardens is an ecologically sound practice, because they are often better suited to local conditions. They can also provide important shelter for wildlife in your garden. In my backyard is a large native Mock-orange, along with many other plants that have traditionally provided cultural materials: Oceanspray, Red Osier Dogwood, Sword Fern and Fireweed. The Mock-orange has particularly attractive and fragrant flowers, and hidden among its leafy branches is a robin's nest that sheltered at least one family last summer. Arthur Kruckeberg's *Gardening with Native Plants of the Pacific Northwest* (1996) is an ideal reference to propagation and growing of many native plants.

Many new botanical publications relating to British Columbia have been written in the past two decades. You can find these in the revised References section at the back of this book. They include a series of plant guides published by Lone Pine (MacKinnon et al. 1992, Pojar and MacKinnon 1994, Johnson et al. 1995, Parish et al. 1996 and Kershaw et al. 1998), *Ecosystems of British Columbia* (Meidinger and Pojar 1991) and a revised and updated version of Ches Lyons' classic *Trees and Shrubs to Know in British Columbia* (Lyons and Merilees 1995). They also include other general floras: *Vascular Plants of British Columbia*, a four-volume set by Douglas et al. (1989–94), a new and detailed *Trees and Shrubs of British Columbia* (Brayshaw 1996b), and three monographs on particular plant groups by T.C. Brayshaw (1985, 1989, 1996a).

In addition to the people mentioned in the preface to the first edition, many more have continued to contribute much to a better understanding and appreciation of traditional plant materials and to the philosophies surrounding their sustainable use in British Columbia: Mary Thomas, Nellie Taylor and Aimee August of the Secwepemc Nation; Mabel Joe, Annie York, Hilda Austin and Louie Phillips of the Nlaka'pamux Nation; Alec Peters, Margaret Lester and Nellie Peters of the Lil'wat Nation; Bill Edwards, Desmond Peters Sr and Edith O'Donaghey of the Stl'atl'imx Nation; Mac Squinas, Maddie Jack and other elders of the Ulkatcho Nation; Lena Jumbo, Gloria Frank, Arlene Paul, Stanley Sam, Roy Haiyupis, Chief Earl Maquinna George and Dr Richard Atleo (Chief Umeek) of the Ahousaht Nation; Larry Paul and Alice Paul of the Hesquiaht Nation; Ida Jones, Chief Charlie Jones and John Thomas of the Pacheedaht Nation; Elsie Claxton and all of her family of the Saanich Nation; Violet Williams of the Hulq'umi'num and Saanich nations; Arvid Charlie of the Hulq'umi'num Nation; Chief Adam Dick (Kwaxistala), Daisy Sewid-Smith (Mayanilth) and Kim Recalma-Clutesi of the Kwakwaka'wakw Nation; Barbara Wilson, Captain Gold and other members of the Haida Gwaii Watchmen Program of the Haida Nations; Judy Thompson of the Tahltan and Tsimshian nations; Sarah and Loveman Nole, and Julia and Charles Callbreath of the Tahltan Nation; Helena Myers and Linda Myers of the Tsilhqot'in Nation. Many of these people are featured and acknowledged in the publications cited, and there are many other people not mentioned here.

I would like to thank George Nicholas for reading the new preface, especially the part on archaeology. Thanks also to Gerry Truscott, publisher at the Royal British Columbia Museum and the production staff at the University of British Columbia Press for their work in producing this second edition. Thanks also to John Veillette of Anthropological Collections at the Museum for his advice and assistance with artifact photography. As always, I am grateful to my husband, Bob Turner, for his many photographs appearing in this edition and for his editorial counsel. Most of all, I thank and acknowledge the many authorities of the First Nations on the use of plant materials and artifacts, both past and present, whose memories, experiences and expertise are recorded in this book. Many of these people are named in the prefaces to both editions.

I hope that this book will be useful and informative as a reference guide, highlighting a group of utilitarian plants, often little known and underappreciated today. Yet, these plants, used for clothing, fuel, construction materials, ceremonial regalia, fishing gear, containers for transport and storage, and in countless other ways, supported healthy and productive lifestyles for peoples all over the province for thousands of years. Many are still important today and their importance will persist for countless generations to come.

Nancy J. Turner, Ph.D., F.L.S.
School of Environmental Studies
University of Victoria
Research Affiliate
Royal British Columbia Museum
June 1998

Hand-made plant-products, clockwise from bottom left:
Yew-wood snowshoes strung with rawhide, made by Alec Peters (Lower
Stl'atl'imx); willow-bark doll made by Mary Thomas (Secwepemc); woven
cedar-bark basket (Kwakwa̱ka̱'wakw); birch-bark basket with a rim of
cedar root and cherry bark, made by Mary Thomas; birch-wood spoon
(Gitxsan); and Slough Sedge basket (Nuu-chah-nulth).

INTRODUCTION

The importance of plant foods and medicines to First Peoples is well appreciated, but the role of plants in aboriginal material cultures – in many cases just as vital – is often overlooked. In British Columbia, where the winters are long and cold, where most foods must be cooked to make them palatable, where waterproof clothing and watertight shelters are indispensable, where the most efficient means of transportation in the early days was often by water, where the aboriginal hunting/fishing/gathering economy required a myriad of special tools, and where metal of any type was relatively rare and pottery virtually unheard of, plant materials were not only useful but absolutely essential to survival.

Heat, shelter, transportation, clothing, implements, nets, ropes and containers – necessities of life in the northwestern North American environment – all were afforded by the great abundance and variety of plants in the area. First Peoples also used plants in many other ways, such as for decoration and ornamentation, as scents, cleansing agents and insect repellents, and in recreational activities.

Over the millennia, the First Peoples of British Columbia have become highly skilled in the arts of working with plant materials. The cedar-wood canoes, totem poles and kerfed boxes of the coastal groups, especially those of the central and northern regions, are thought by many to be the most superb examples of woodworking craftsmanship to be found anywhere, and the coiled split-cedar-root baskets of the Interior Salish peoples, with their intricate imbricated designs, are world famous. There are innumerable other equally impressive examples of proficiency in the utilization of plants and plant products.

The purpose of this handbook is to provide information on the many different types of plants used as materials by the First Peoples of British Columbia and some adjacent territories; at the same time, it is to describe how these plants were employed and by which cultural and linguistic groups. Hopefully, this book will be of interest to many different segments of the population: to aboriginal people seeking information about their past cultures; to professionals, such as archaeologists, ethnologists and botanists, investigating certain aspects of material culture, identifying artifact materials or seeking information about the practical applications of plants; and to members of the general public. I have kept all these groups in mind during the planning and writing of this book, and I have tried to be as complete and detailed as possible, while keeping technical language to a minimum. Throughout the book are photographs of the plants to aid in their identification and of some articles made from plant materials to give the reader a better understanding of the practical aspects of plant use.

For those who might wish to employ some of the plants mentioned in this book for carving, weaving or some other form of manufacture, the possibilities are limitless. But please understand that almost any large-scale harvesting of plant materials – bark, roots, wood or stems – is potentially detrimental to the plants and may affect their survival. With herbaceous perennials, such as Stinging Nettle, Indian Hemp, Tule or Cattail, the gathering of stems or leaves is not likely to deplete the populations to any measurable degree, since these plants tend to grow in large patches and the rootstocks will grow new stems the following year. But many other species, especially trees, do not regenerate as easily.

If you remove the bark around the entire circumference of a tree, you will kill the tree. Even if you only strip off part of the bark, you will still injure the tree, leaving the wood exposed to insect and fungal infestations and decreasing the tree's capacity to transport water and nutrients from the roots. Harvesting wood is even more detrimental, especially if you cut down an entire tree. Some species, such as Western Yew (especially the larger individuals), have become rare in some places. This is due at least in part to clearcutting, habitat destruction and overzealous harvesting by wood-carvers. More recently, Western Yew has been put under more pressure from bark collection for an anti-cancer medicine.

If you are planning to use plant materials, please consider carefully the consequences of harvesting, and practise the utmost care and discretion in removing any plants or parts of plants from any natural area. First Peoples have been vigilant managers of their resources, including plant materials. They remove cedar bark in single or double strips, leaving most of the bark intact so that the tree remains alive. They carefully select the trees to be cut down; sometimes they split planks from a standing cedar tree instead of felling it. Today, trees that show evidence of partial harvesting are called Culturally Modified Trees. They provide important documentation of human use and occupation of a given landscape.

First Peoples cut from living shrubs and trees the withes and woody stems they used for fishing weirs and traps, baskets, sweat-lodge frames, and bows and arrows. They selectively harvested tree roots for baskets and lashing in a way that resembled the pruning of branches. And they carefully harvested the leaves and stalks of herbaceous perennials: some basket makers say that these plants are more productive when they are tended and routinely harvested.

Format

The main body of this book lists the plants used by British Columbia's First Peoples in a technological capacity. Each listing contains a botanical description, habitat, distribution of the plant in the province and a discussion of the ways in which aboriginal peoples used it.

The plants are arranged in a partially botanical and partially alphabetical order. The major plant groups are in their traditional order: algae, lichens, fungi, mosses, ferns and their relatives, conifers, and flowering plants. The flowering plants are further divided into two subgroups: monocotyledons and dicotyledons. The ferns, conifers, monocotyledons and dicotyledons are classed in family groups, which are presented in alphabetical order of the scientific family name. The sections on algae, lichens, fungi and mosses are too small for them to be subdivided into families. Whether in family or major group, the species themselves are listed in alphabetical order of their scientific names.

The scientific names of species and families appear on the right-hand side of the page and the common names on the left-hand side. The most widely accepted common name is featured in the heading and used throughout the text, but alternative common names (if any) are also given. Locally used names, which have limited application and can sometimes be confusing, appear in quotation marks.

For convenience, I have used common names throughout the text, except when a plant has no recognized common name or when it is clearer to use the scientific name. Common names of species are capitalized to make them easily recognizable. Plants mentioned but not described in this book are listed with their scientific names in Appendix 2; animals are also listed in this appendix.

For algae, lichens, fungi, mosses, ferns and grasses, I discuss the uses of the entire group in general, then give detailed treatments of the species of greatest technological importance. Many plants played minor roles in British Columbia aboriginal technology, being used casually or by only one group of people – these are listed in Appendix 1.

Vegetation and aboriginal cultures do not change at political borders. Thus, many of the plants discussed in this book were used in similar ways by the First Peoples of neighbouring areas – the Tlingit of Alaska; the Makah, Quileute and various Coast Salish groups of western Washington; the Okanogan and other Interior Salish groups of central and eastern Washington; the Sahaptin of the mid-Columbia; the Flathead and Ktunaxa (Kutenai) of Montana; and the Blackfoot of Alberta, to name a few. I have included information on the material uses of plants by these peoples where available and when it contributed to the discussions of British Columbia aboriginal plant technology.

Sources of Information

I consulted many published references, both botanical and ethnographic, in the preparation of this book. The botanical descriptions and the notations on the habitat and distribution of plant species are partly based on personal observation, but mainly on various botanical works. Foremost among these was the five-volume flora, *Vascular Plants of the Pacific Northwest* (Hitchcock et al. 1955-69). Also helpful were the handbooks of the Royal British Columbia Museum on the various plant families and other groups.

Elder members of First Nations communities – many of whom are named in the prefaces – contributed largely to the information on the uses of plants by aboriginal peoples. Their knowledge came from personal experiences and from past observations and conversations with the elders of previous generations. As noted in the Preface to the Second Edition, there are still many people, as of the late 1990s, skilled in the use of plant materials.

Some of the most significant published sources for First Peoples' uses of plant materials are: Franz Boas's *Ethnology of the Kwakiutl* and his other books (mainly on Kwakwaka'wakw culture), James Teit's books on Interior Salish cultures and Rev. A.G. Morice's writings on Athapaskan cultures (e.g., Morice 1895).

Randy Bouchard and Dorothy Kennedy of the British Columbia Indian Language Project, provided considerable information from their own field research on plant materials used by various Salishan groups of the coast and the interior. They also compiled a number of excellent papers on the use of fish and marine life by Salishan groups; these include discussions of plant materials used in fishing technologies. I obtained information on plant technology of the First Peoples of western Washington from *The Ethnobotany of Western Washington* (Gunther 1945), of the Flathead and Ktunaxa peoples of western Montana from *Plant Taxonomy of the Salish and Kootenai Indians of Western Montana* (Hart 1974), and of the Blackfoot of Alberta from "Blackfoot Indian Utilization of the Flora of the Northwestern Great Plains" (Johnston 1970) and *Ethnobotany of the Blackfoot Indians* (Hellson and Gadd 1974).

These sources and many others are listed in References.

The Physical Environment

No province in Canada has greater topographic, climatic and biological diversity than British Columbia. From the rugged coastline on the Pacific to the steep, jagged peaks of the Rocky Mountains in the east, the landscape is a continuously changing series of hills, mountains, plateaus, plains, valleys, canyons, marshes, muskegs, lakes and rivers. The annual precipitation of 250 cm or more on the west coast of Vancouver Island contrasts sharply with the semi-desert conditions in parts of the southern interior, where combined rain and snow barely exceeds 25 cm a year. The mild, Mediterranean-like climate of southeastern Vancouver Island, which can support blooming flowers even in mid winter, is far removed from the chilling subarctic temperatures of the central and northern interior, where the mean temperature for the month of January can reach −18°C.

In a region characterized by so much variety, it is difficult to summarize the essential geographical and biological features in a few pages. But it is the very diversity of landscape, climate and vegetation that has allowed First Peoples to use so many different plants in such a variety of ways. Thus, it is important to understand the environment of the region covered by this book.

The two predominating physiographic features of the province, enclosing all but the northeastern corner, are the Coast and Cascade mountain ranges on the west and the Rocky Mountains on the east. All are part of the great system of mountains known as the Cordillera, which extends north to south along the entire western side of North and

South America. The northeastern corner of the province, east of the Rockies, is part of the great Interior Plains region.

Between the Coast and Rocky mountain ranges is a system of alternating mountains and plateaus. In the north, the dominant plateau is the Stikine Plateau, which is surrounded by the Omineca, Hazelton, Skeena and Cassiar ranges. In the south, the extensive Interior Plateau lies between the Columbia Mountains and the Coast Mountains. The Columbia Mountains are made up of four smaller ranges: the Purcell, Selkirk and Monashee mountains in the south, and the Cariboo Mountains to the north.

This mountain-plateau system contains a multitude of lakes and rivers with their associated valleys, flood plains, marshes, canyons, gorges and benchlands, separating ranges and cutting through the plateaus. Immediately west of the Rocky Mountains is a long, straight valley, known as the Rocky Mountain Trench. Varying between 3 and 16 kilometres wide, the trench flanks the Rockies south to north. Most of the rivers west of the Rocky Mountains flow southward and westward to the Pacific Ocean, transecting the Coast Range. The land east of the Rockies drains northward into the Arctic Ocean.

The Coast Range extends onto Vancouver Island and Haida Gwaii. The coastline of these islands and the mainland is, for the most part, rocky and indented with a multitude of steep, narrow fiords and inlets. But there are some shallow bays and wide stretches of sandy beach. Lowland coastal plains occur along the east coast of Vancouver Island and the northeastern corner of Graham Island in Haida Gwaii. At the estuaries of the major rivers, such as the Fraser, are fertile alluvial flood plains and deltas.

Few First Peoples resided for long periods in the mountains of British Columbia. For their major habitations, coastal peoples preferred the bays, inlets and estuaries, and interior peoples inhabited the plateaus, benchlands and river valleys. The ocean and the interior waterways were primary sources of food and major routes for transportation. First Peoples ventured into the uplands to hunt and gather food, and traversed them on occasional trading expeditions; they also went on spiritual journeys into the mountains. But the steep, rugged slopes and snow-covered peaks were major barriers to travel and communication and must certainly account, in part, for the great linguistic and cultural diversity among the First Peoples of British Columbia. Despite the cultural differences, the various aboriginal technologies were by-and-large those of lowland water-oriented societies. Canoes – dugout on the coast, and birch-bark or dugout in the interior – were important pieces of equipment, as were the various implements required for fishing and marine hunting: paddles, nets, lines, weirs, harpoons and spears. Thus, it can be said that the very nature of the landscape had some effect in determining how plants were used in aboriginal technologies. The multitude of baskets, bags, boxes and other containers people made and used certainly reflected their lifestyles, with transporting and storing goods a major requirement for survival.

British Columbia lies in the North Temperate Climatic Zone. The weather patterns are largely controlled by air masses and pressure systems moving across the province from west to east. But these basic patterns are modified considerably by topography, altitude and latitude, thus no two localities in the province have exactly the same climatic conditions.

The climate in different areas of the province had a significant effect, both direct and indirect, on the types of plants First Peoples used in their technologies and the ways in which they employed these plants. For example, climatic patterns influenced the architectural styles and materials used for dwellings and the types of clothing worn in the various regions. In the rainy coastal zone, the large cedar-board houses and woven cedar-bark clothing were eminently suitable; in the interior, semi-subterranean pit houses and skin clothing were more appropriate for the severe winter weather, while temporary tipis (or mat huts) and plant-fibre clothing served in the hot summer months, although some of the more nomadic peoples used skin tipis for shelter throughout the year. Various items, such as snowshoes, widely used in the interior, were also directly related to climate.

In keeping with the geographic and climatic diversity of British Columbia, the vegetation also varies greatly, both locally and regionally. Biologists divide the province into 14 biogeoclimatic zones. As the name suggests, the zones are distinguished by their biology (mostly vegetation), geography and climate. There are three zones on the coast and ten in the interior; the fourteenth, the Alpine Tundra Zone, occurs throughout the province.

The westernmost biogeoclimatic zone is the Coastal Western Hemlock Zone, which occupies Vancouver Island, Haida Gwaii and the entire mainland coast below a range of about 1,000 metres in the south to 300 metres in the north; in the Strait of Georgia, this zone lies above the Coastal Douglas-fir Zone (see next page). With an average annual precipitation of 165 to 665 cm, the Coastal Western Hemlock Zone supports Western Red-cedar, Sitka Spruce, Amabilis Fir and Red Alder, as well as Western Hemlock.

The Mountain Hemlock Zone occupies the subalpine elevations, above the Coastal Western Hemlock Zone (from about 900 to 1,700 metres in the south, and 300 to 600 metres on the Alaskan Panhandle). The Mountain Hemlock Zone has an average annual precipitation of 178 to 432 cm, much of it in the form of snow. Trees associated with this zone are Mountain Hemlock, Yellow-cedar, Amabilis Fir and, occasionally, Subalpine Fir.

The Coastal Douglas-fir Zone is confined to the leeward side of southern Vancouver Island, the Gulf Islands and the lowland areas of the adjacent mainland. The average precipitation in this zone is 66 to 152 cm, and it is characterized by the presence of the coastal variety of Douglas-fir (var. *menziesii*) and a number of other tree species – Arbutus, Garry Oak, Lodgepole Pine, Grand Fir, Broad-leaved Maple and Red Alder.

Two zones occupy the warmest and driest river valleys of the southern interior. The Bunchgrass Zone is limited to the lower elevations of these valleys. It is dominated by Bluebunch Wheatgrass and does not support the growth of trees. Typical shrubs are Big Sagebrush, Waxy Currant and Antelope-brush. Bordering much of the Bunchgrass Zone's upper limits is the Ponderosa Pine Zone. As the name indicates, Ponderosa Pine is the dominant tree, although the interior variety of Douglas-fir (var. *glauca*) is common in cooler, moister areas.

The Interior Douglas-fir Zone is also dry, with a total annual precipitation of 36 to 56 cm. Besides Douglas-fir itself, some common trees of this zone are Ponderosa Pine, White Pine, Lodgepole Pine, Western Larch, Trembling Aspen, Black Cottonwood, Rocky Mountain Maple and Paper Birch.

The Montane Spruce Zone is characterized by Engelmann and hybrid spruces, as well as some Subalpine Fir. Due to past fires, successional forests of Lodgepole Pine, Douglas-fir and Trembling Aspen dominate the landscape. This zone borders the upper limits of the Interior Douglas-fir Zone.

The Interior Cedar – Hemlock Zone occurs in lower and middle elevations in the interior wet belt and in the Skeena and Nass river valleys in the northwest. Cool, wet winters and warm, dry summers make this zone the most productive in the interior, producing the widest variety of coniferous trees. Western Red-cedar and Western Hemlock are characteristic, but spruces and Subalpine Fir are also common, as well as Douglas-fir and Lodgepole Pine in the drier areas.

One of the wettest zones in the interior is the Engelmann Spruce – Subalpine Fir Zone, with an annual precipitation of 41 to 183 cm. This is a montane zone, 1,200 to 2,200 metres elevation in the south and 1,000 to 1,600 metres in the north, so most of its precipitation is snow. Common trees in this zone, besides Engelmann Spruce and Subalpine Fir, are Lodgepole Pine, Whitebark Pine and Alpine Larch. All of these species tolerate harsh winters with extended periods of frozen ground.

The Sub-boreal Spruce Zone is in the central interior, mostly on plateaus. It is an intermediate zone between the southern Douglas-fir forests and northern boreal forests. The dominant trees are hybrid Engelmann-White Spruce and Subalpine Fir. Wetlands in poorly drained areas are common here.

The Sub-boreal Pine – Spruce Zone is on the high plateau of the central interior in the rain shadow of the Coast Mountains. The cold, dry climate and long winters make this zone low in productivity. Extensive forest fires have encouraged the growth of Lodgepole Pine in many areas, and some White Spruce also grows here. Pinegrass and Kinnikinnick are common understorey plants.

The Boreal White and Black Spruce Zone is part of the belt of boreal coniferous forest that stretches across northern Canada. In British Columbia it occupies the Great Plains east of the Rocky Mountains and river valleys in the northwest. Long, cold winters and the short growing season make for poor plant growth. Black Spruce bogs and stands of White Spruce and Trembling Aspen are common.

The Spruce–Willow–Birch Zone occupies the subalpine areas in the north. The climate is severe: upper elevations are dominated by shrubs and scrub birch and willow, and lower elevations by open forests of White Spruce and Subalpine Fir.

Finally, the Alpine Tundra Zone occurs throughout the province above the timberline. Conditions on the high mountains are too severe for woody plants, except in dwarf forms. The vegetation is dominated by herbs, mosses, lichens and some dwarf shrubs. Aboriginal peoples occasionally ventured into this zone to hunt animals or gather plants.

You can learn more about these zones and their vegetation from the Ministry of Forests' wall map, *Biogeoclimatic Zones of British Columbia* (Forests 1992), and from *Ecosystems of British Columbia* (Meidinger and Pojar 1991).

Many of the plants mentioned in this book are restricted to one or two biogeoclimatic zones, or at least grow more abundantly in some zones than in others. Hence, some First Peoples had better access than others to certain species. Those living on the coast and in areas of the southern interior where Western Red-cedar was abundant were fortunate indeed. But interior peoples had access to other species, such as

Paper Birch, not widely available on the coast. The ingenuity and inventiveness of the people in adapting the plant materials at hand to suit their requirements is remarkable. The bent cedar-board boxes of the coastal peoples and the sewn birch-bark vessels of the interior groups served the same general function, but the materials themselves and the ways in which they were handled were strikingly different.

Trading overcame some of the discrepancies in available plant materials; but, in general, First Peoples were limited to the plants that grew in their particular areas. In every case they did a superb job in using plants to their fullest advantage.

The fauna of British Columbia – the birds, mammals, reptiles, amphibians, fish and invertebrates – cannot be discussed here in detail, but it must be mentioned that the animals are as diverse as the plants. Their distribution is largely affected by climate, geography and vegetation, and their influence on aboriginal lifestyles was immense. Animals provided the greatest source of food, at least in terms of quantity, and as such, they had a profound effect on the types of implements and equipment used by First Peoples. Many of the tools made from plant products – such as bows and arrows, harpoons, spears, and nets – were directly related to the quest for animal foods, and many others – such as canoes, grease boxes and drying racks – were partly connected with hunting and fishing activities. In most areas, animal products were not as important technologically as plant products, except in the northern interior, where vegetative diversity is lower. But throughout the province, people used skins, sinew, bone, antler, horn, animal fat, feathers and Porcupine quills to some extent.

Before Europeans arrived here, the plants and animals of the province provided sustenance for hundreds of thousands people (R.T. Boyd 1990). In their given environment, these people had highly specialized, superbly adapted technologies – a credit to their skill, knowledge and ingenuity.

The First Peoples of British Columbia and Adjacent Areas

In this book, I have distinguished the First Peoples of British Columbia by the languages they speak. The table on page 26 shows the name of each language group, its general pronunciation and its former or alternative name, where applicable. There are differences in the ways individual First Peoples say these names, so the pronunciations are, at best, approximations. The territories occupied by these groups are shown on the map on page 27.

More than 30 languages are spoken in the region covered by this book; some are closely related, while others are as remote from each other as English and Swahili. The languages spoken by the southern coastal peoples – Straits Salish, Halkomelem, Squamish, Sechelt, Comox and Nuxalk – form part of the coastal division of the Salish Language Family, while those spoken by the south-central interior peoples – Nlaka'pamux, Stl'atl'imx, Okanagan and Secwepemc – form part of the interior division of the Salish Language Family. On the coast, the languages spoken by the Kwakwaka'wakw, Heiltsuk, Haisla, Oweekeno and Nuu-chah-nulth belong to the Wakashan Language Family. The languages spoken by the peoples of the central-northern interior –Tsilhqot'in, Carrier (Dakelh), Cree, Wet'suwet'en, Sekani, Dunne-za, Dene-thah, Tahltan, Kaska and Tagish – form part of the Athapaskan Language Family. This family is in turn considered to be broadly related to the Tlingit language of the British Columbia interior and coastal Alaska. The language of the Haida, on Haida Gwaii and some of the Alaskan coastal islands, is considered a linguistic isolate, according to the most recent research. The languages of the Tsimshian, Nisga'a and Gitxsan peoples form part of the Penutian Language Family. In southeastern British Columbia, the language of the Ktunaxa peoples is believed to be, like Haida, a linguistic isolate.

Whereas each aboriginal group is unique, many of them, even when completely unrelated linguistically, share common cultural traits, and can be categorized at a general level into major cultural units. All of the coastal peoples, from the Straits Salish of southern Vancouver Island to the Haida and Tsimshian in the north, belong to the Northwest Coast Culture Area, which also encompasses the coastal peoples of Washington, Oregon and northern California, as well as the coastal Tlingit of Alaska. The Northwest Coast Culture Area is characterized by a number of distinctive features, including a marine fishing and water-oriented economy, with special dependence on the Pacific salmon, and extensive use of the Western Red-cedar for construction of dugout canoes, plank houses and other wooden objects. The bark fibre of both Red- and Yellow-cedar was commonly employed for making clothing, ropes, blankets and mats.

Interior First Peoples belong to two main cultural units. The Plateau Culture Area includes the Salishan peoples of the Interior Plateau and the Ktunaxa of eastern British Columbia, along with the various Salishan peoples and the Nez Percés and Sahaptin peoples of central

First Nations of British Columbia

Language Family
  People
    Dialect
	Pronunciation*	Former or
		Alternate Name

Interior Salish	Say-lish	
Stl'atl'imx	Stlat-liem*x*	Lillooet
Upper		Fraser River
Lower		Lil'wat
Okanagan	O-kan-a-gan	
Northern		
Southern		
Secwepemc	Se-wep-m*x*	Shuswap
Nlaka'pamux	Ng-khla-kap-muh*x*	Thompson
Upper		
Lower		
Ktunaxa		
Ktunaxa	Doon-ah-hah	Kootenay, Kutenai
Athapaskan		
Tsilhqot'in	Tsil-ko-teen	Chilcotin
Carrier (Dakelh)	(Da-kelh)	
Ulkatcho	Ul-gat-cho	Carrier
Cree	Kree	
Wet'suwet'en	Wet-so-wet-en	Babine Carrier
Sekani	Sik-an-ee	
Dunne-za	De-nay-za	Beaver
Dene-thah	De-nay-ta	Slave(y)
Tahltan	Tall-ten	
Kaska	Kas-ka	
Tagish	Ta-gish	
Tlingit	Tling-git	
Inland Tlingit		
Tsimshian	Sim-she-an	
Nisga'a	Nis-gaa	Nishga
Gitxsan	Git-ksan	Tsimshian

Coastal Groups		
Haida	Hydah	
Wakashan	Wak-a-shan	
Haisla	Hyzlah	Kitimat
Heiltsuk	Hel-tsuk	Bella Bella
Oweekeno	O-wik-en-o	Kwakiutl
Kwakwaka'wakw	Kwak-wak-ee-wak	Kwakiutl, Kwagiulth
Nuu-chah-nulth	New-chah-nulth	Nootka
Ditidaht	Dit-i-daht	Nitinat
Coast Salish	Say-lish	
Nuxalk	New-halk	Bella Coola
Comox	Koe-moks	
Sechelt	Seeshelt	
Squamish	Skwamish	
Halkomelem	Halk-o-may-lem	
Straits Salish		

*The pronunciations are approximations. Note that the italic *x* in the pronunciation column designates the fricative sound similar to the German *ich*.

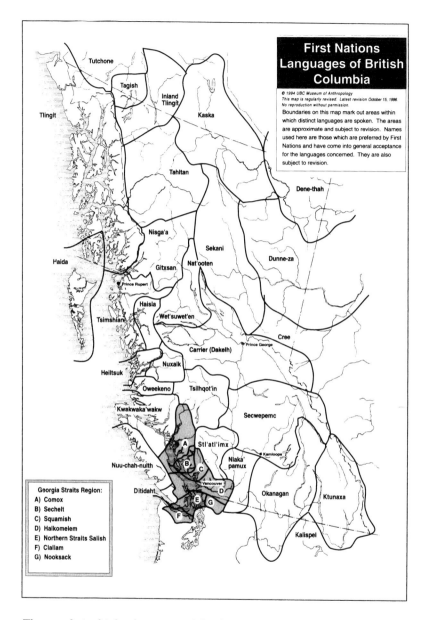

First Nations Languages of British Columbia

© 1994 UBC Museum of Anthropology
This map is regularly revised. Latest revision October 15, 1996.
No reproduction without permission.
Boundaries on this map mark out areas within which distinct languages are spoken. The areas are approximate and subject to revision. Names used here are those which are preferred by First Nations and have come into general acceptance for the languages concerned. They are also subject to revision.

Georgia Straits Region:
A) Comox
B) Sechelt
C) Squamish
D) Halkomelem
E) Northern Straits Salish
F) Clallam
G) Nooksack

The people in this book are named for the languages they speak; but Kwakwaka'wakw is the name of the people who speak the Kwakwala language. For consistency, I have used the name of the people - not their language - on this map.

and eastern Washington, Idaho and western Montana. These groups, especially the Ktunaxa, possessed many of the cultural traits of the Plains First Peoples, but were generally less nomadic. A major food of the Plateau Culture Area peoples was – and still is – Pacific salmon, which came in great numbers every year up the Fraser and Columbia river systems to spawn. Edible plants, especially root vegetables, and game are also important traditional foods. People used a great variety of plant materials, including some plant fibre for clothing, Indian Hemp for rope and line, and split cedar root and birch bark for basketry.

The Subarctic Culture Area includes the Athapaskan groups of central and northern British Columbia along with the peoples of central Alaska, the Yukon, the western Northwest Territories, and northern Alberta and Saskatchewan. The peoples in this region were widely scattered and semi-nomadic, travelling by birch-bark or spruce-bark canoe, or in winter, by snowshoes, and depending heavily on the meat of large game animals such as Moose and Caribou. They used some plant materials, such as birch bark, but made most of their clothing and shelter from animal products.

The interior groups bordering the coastal area – notably the Lower Stl'atl'imx, Lower Nlaka'pamux, Inland Tlingit, Gitxsan and Nisga'a (whose territory partially borders the ocean) – are culturally transitional, sharing many of the cultural features of the Northwest Coast Culture Area. They used many typically interior plant products, but they also had access to many coastal plant materials not readily available to other interior groups.

You can find more information about the characteristic traits of these Culture Areas in the Smithsonian Institution handbooks on *Northwest Coast* (Suttles 1990), *Plateau* (Walker 1998) and *Subarctic* (Helm 1981).

It is important, when observing and comparing the use of plant materials in different regions of the province, to understand cultural differences as well as vegetational ones. As a general rule, First Peoples used a plant of major technological importance throughout its area of distribution. Western Red-cedar, Western Yew, Paper Birch, Indian Hemp and Red Alder were employed almost universally within their ranges. There were exceptions to the rule, of course. For example, peoples of the southern coast and the interior valued Oceanspray – the so-called "Ironwood" – for its hard, strong wood, using it mainly to make digging sticks; but the Nuxalk did not appear to use it much, even though it grows in their area. Thus, sometimes the use of plants and the ways in which they were prepared for use seem as much related to a group's culture as to the plant's availability.

Plant Materials

The First Peoples of British Columbia have employed plants in many different capacities. From woods they have carved implements, containers and canoes, built houses and shelters, and fuelled fires for heating homes and for cooking and smoking foods. They have used sheets of bark to make containers and canoes, for lining underground food caches, and as roofing. With bark, stem, leaf and root fibres they have made twine, ropes, fishing lines and nets, baskets, bags, mats, and clothing. They have also used plants and plant products as bedding and floor coverings, as lining for berry baskets, drying racks and steaming pits, to make storage vessels, water conductors and Soapberry beaters, as herring-spawn collectors, infant diapers, abrasives and tinder to start fires, for dressing wounds, and to make paints, dyes, tanning agents, glues, animal poisons, insect repellents, scents, soaps and cleansing agents, decorations, and toys. They have even used plants as biological indicators for various seasonal and climatological events.

In this section, I offer a few examples of the ways First Peoples used – and still use – plants in their technology. While the examples are not exhaustive, I hope it will at least give you a feeling for the great versatility of the vegetation available in British Columbia and for the ingenuity of aboriginal peoples in adapting the plants for their use.

Few woods could match the versatility of Western Red-cedar. Soft, easy to split and work, yet durable and rot-resistant, this wood was highly valued throughout its range, especially on the coast where it grows abundantly and to tremendous heights and girths. Red-cedar provided the wood for house planks and posts, canoes, totem poles and innumerable smaller items. It was also an excellent fuel, burning quickly with a hot flame. And the flexible young branches (withes) made excellent ropes and lines. Cottonwood has some of the same characteristics as cedar, and interior peoples often used it to make dugout canoes and as fuel. They also valued the dead wood and dried roots of Cottonwood for making the hearth and drill to start friction fires.

First Peoples throughout the province used the woods of many other trees and shrubs, most in more-or-less specialized ways. They used the hard, resilient wood of Western Yew to make bows, wedges, digging sticks, snowshoe frames, adze handles, and shafts for harpoons and spears; many considered Wild Crabapple a good substitute for Yew in making these implements. They also considered Yellow-cedar, Rocky Mountain Juniper, Flowering Dogwood and Rocky Mountain Maple

Elsie Claxton of Saanich examines a cedar bough to assess its quality as a basket material.

good woods for making bows, and Yellow-cedar good for paddles. People frequently used the hardwood shrubs Oceanspray and Mock-orange to make digging sticks and smaller items, such as combs, mat-making needles and arrows. Throughout the interior, people used Saskatoon Berry wood to make arrows, drying racks, and frames for sweathouses, canoes and baskets; they also used Red-osier Dogwood branches to make frames and basket rims. The saplings of Vine Maple and Rocky Mountain Maple, like Yew saplings, made fine snowshoe frames, while Broad-leaved Maple was good for spindle whorls and paddles. Carvers, especially along the coast, made masks and dishes of all sizes from Red Alder, a soft, light-coloured, even-grained wood; and many people valued alder as a fuel, especially for smoking meat and fish. Birch was also considered a good wood for carving. People made poles for many purposes from the trunks of Lodgepole Pine and Western Hemlock, and used Ponderosa Pine in construction and, occasionally in the southern interior, for dugout canoes. Some people used Lodgepole Pine, Ponderosa Pine or Pacific Willow as the hearth and drill for making friction fires. Coastal peoples moulded the knots and roots of Douglas-fir, Western Hemlock, Amabilis Fir and Sitka Spruce into cod and halibut hooks, and interior peoples used Douglas-fir branches and saplings to make dipnet hoops and handles.

Tree barks were almost as useful as wood. In many areas, Douglas-fir bark had the reputation of being an excellent fuel. Almost everyone is familiar with the widespread use of birch bark for making canoes and baskets, not only in the British Columbia interior, but all across the country. Other barks, including Engelmann Spruce, Subalpine Fir, Grand Fir and White Pine, could be used similarly. People made storage buckets and lined underground food caches with Cottonwood bark, and some coastal groups used sheets of Red-cedar bark to make buckets and to roof their houses and shelters. The fibrous inner bark of both Red-cedar and Yellow-cedar could be split into thin strands and used to make capes, skirts, bodices, hats, mats, blankets, baskets, rope and twine; shredded inner bark was good for baby diapers, napkins, towelling and tinder. The tough, waterproof bark of the Bitter Cherry

was ideal for wrapping the joints of implements, particularly those to be used in or
near water, such as harpoons, fish-spears and
fish-hooks. Basket makers also used this bark
to imbricate patterns on coiled baskets – it
has a beautiful varnished-red colour and
stains easily to jet black. First Peoples of the
province, particularly those in the interior,
used several other bark fibres for making
ropes, bags and clothing; these included
several willow species, Silverberry, White
Clematis and Big Sagebrush.

Gitxsan birch-wood
spoon.

First Peoples also employed other plant fibres. To make twine, fishing lines and nets,
they spun stem fibres of herbaceous species
such as Indian Hemp, Spreading Dogbane,
Stinging Nettle and Fireweed. Most coastal peoples used the strong, pliable stems of Bull Kelp to make fishing and anchor lines. People
throughout the province wove baskets, bags and mats with the fibrous
leaves of grasses and grasslike plants, including Common Reed Grass,
Giant Wildrye, Bear-grass (actually a lily), surf-grasses, sedges and Cattail, as well as the stems of Tule; they also used them to decorate articles woven from other materials. Fibrous roots, such as those of
Red-cedar, Sitka Spruce and Engelmann Spruce were split lengthwise
into thin strands and woven into baskets, hats and other articles, and
used for binding, tying and sewing. Weavers sometimes spun different
kinds of fibres together or used them in weaving or sewing different
parts of a basket or other item.

Aboriginal people commonly slept on Cattail or Tule mats, or
boughs of conifers, such as Douglas-fir, Subalpine Fir, Amabilis Fir,
Grand Fir, Lodgepole Pine and even spruce (as long as the prickly
needles were not next to the skin). They also stuffed cloth mattresses
and pillows with the woolly seed-fluff of Cattail, Cottonwood or
Fireweed. To cover the floors of summer and winter houses and sweat-
lodges, people spread out Sword Fern and Bracken Fern fronds,
grasses, willow and Red-osier Dogwood branches, and conifer boughs.
To line their steaming pits, they used the types of vegetation that would
provide moisture and keep the food clean, such as Skunk Cabbage
leaves, the branches of Salal, wild roses or Red-osier Dogwood, seaweeds, mosses, Bluebunch Wheatgrass, Fireweed, and Shrubby
Penstemon; many of these added flavour to the food. The large, flat

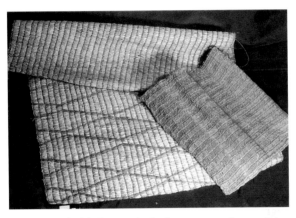

Fibre mats (Nlaka'pamux): the large one underneath is
made of Indian Hemp fibre and Silverberry bark
(RBCM 2581); the small one on top is Tule stems and
Indian Hemp fibre (RBCM 7120).

leaves of the Skunk Cabbage were ideal for drying berries on; they
were also good for lining berry baskets and as makeshift berry and
drinking containers. When Skunk Cabbage was not available, the
broad leaves of other plants, such as Thimbleberry, Cow Parsnip and
Broad-leaved Maple, would suffice.

People of the coast moulded Yew-wood bows and wooden fish
hooks in the hollow stipes of Bull Kelp, and also used the stipes as water
conductors in steaming pits and as sound carriers to create special effects during winter dances. They cured the floats and stems of Bull Kelp
and used them as storage containers for grease and oil. In many parts
of the province, First Peoples hollowed out elderberry stems and curled
birch-bark into tubes to make urine conduits in baby carriers, and
sometimes they used the hollow stalks of grass, Cow Parsnip or
Kneeling Angelica as drinking straws and underwater breathing tubes.

Herring eggs were – and still are – a great delicacy for coastal First
Peoples. All along the coast, during the herring spawning season,
people collected eggs by placing boughs of Western Hemlock and Red-
cedar in the ocean. They also used marine plants, including Eel-grass,
surf-grasses and some types of kelp. If they used kelp, they sometimes
ate it along with the eggs; but, for all the other plants, they stripped off
the eggs and discarded the foliage.

Several aboriginal groups in British Columbia used plants in the
preparation of Indian ice cream, a favourite whipped confection made

from Soapberries. The Nlaka'pamux used Pinegrass, both to dry the berries on and to whip them. Some of the other botanical implements used to whip up Indian ice cream were Thimbleberry leaves, Broad-leaved Maple leaves, Salal branches and beaters made from the bark of Rocky Mountain Maple, Broad-leaved Maple or Silverberry.

The use of shredded cedar bark for diapers, napkins and towelling has already been mentioned. Several other plant products valued for their absorbency were used similarly. These include Old Man's Beard lichens (also used to make shoes and clothing in the interior), sphagnum and other mosses, the seed fluff of Cattail, Fireweed and Cottonwood, the inner bark of some willows, and dry, rotten Trembling Aspen wood.

The silicon-impregnated cell walls of the horsetails and Scouring-rush made them ideal as abrasives. People throughout the province used them to smooth and finish wood or stone carvings. Certain plant products were especially suitable as tinder for starting friction fires or for transporting or storing embers in a "slow match": Spiny Wood Fern roots, the inner fibres of Bracken Fern rhizomes, shredded cedar bark, the root fibres of Balsamroot, Big Sagebrush bark, and tissue from bracket fungi and Cinder Conk.

Many plants were used to make dyes, stains and paints. Foremost among these was Red Alder, which yields, depending on the treatment, colours ranging from bright red to orange to dark brown. It was used to dye cedar bark, basket materials and fishing line. Indian Paint Fungus also yields a red pigment, which was used mainly as a face paint. Other red pigments were made from Western Larch gum and Lemonweed roots. Western Hemlock bark was sometimes used as a reddish-brown dye, and was also employed as a tanning agent. Certain berries – such as Blackcap, Wild Raspberry, Salal and Black Twinberry – and the bright red fruits of Strawberry Blite were mashed and used as purple stains. Hazelnut shoots and larkspur flowers were employed in some areas as blue colourings. Yellow was obtained from the inner bark and roots of Oregon Grape and from the bright yellow-green Wolf Lichen – sometimes they were mixed together. Cottonwood bud resin was used as a yellow paint and as a medium for paints of other colours as well. Grass and green algae, in both wood and water, provided green pigments. Charcoal and organic earth were commonly used to produce black; Flowering Dogwood bark and Grand Fir bark also yielded a black dye. Recently, some groups have used water in which rusty nails have been soaking to colour cedar and cherry bark black. Several kinds of tree barks were used as tanning agents. The Nlaka'pamux and

Sechelt used species of bracket fungi, alone or mixed with deer brains, for this purpose. Some groups heated the mucilage from the inside of prickly-pear cactus stems and used it as a fixative for paints and stains.

Pitch from trees such as Cottonwood, Sitka Spruce, Lodgepole Pine, Western Hemlock and Subalpine Fir provided natural glues for sticking parts of implements together, fastening Bitter Cherry bark wrapping in place, fixing poultices and wound dressings, and waterproofing canoes and baskets. Sometimes, people mixed pitch with animal grease and rubbed it on implements to protect them.

People in the southern interior made arrow poisons from the roots of Rocky Mountain Juniper, Lemonweed, Sagebrush Buttercup, False Hellebore or Stinging Nettle. The Okanagan used Chocolate Tips as a fish poison and an insecticide for livestock. For insect repellents, people used Vanilla-leaf, some of the strong-smelling plants in the aster family – notably Big Sagebrush and its relatives in the genus *Artemisia* – and tree pitch. They valued *Artemisia*s and a number of other plants for their aromatic scent and used them in sachets and as incense and fumigants. Some notable aromatic plants are Rocky Mountain Juniper, Grand Fir, Subalpine Fir, Canby's Lovage, Wild Bergamot, Wild Ginger, Common Sweetgrass and "Indian Celery".

The First Peoples of British Columbia made soaps and shampoos from birch leaves, the ashes and inner bark of Cottonwood, Mock-orange leaves and flowers, and White Clematis leaves, among others. They used the fruits or seeds of a number of plants, including Lemonweed, Silverberry, wild roses, cactus and Arbutus, as beads for necklaces and decorations on clothing, especially for women. Some coastal peoples made special ceremonial costumes for shamans and initiates from the boughs of Western Hemlock and Grand Fir.

The aboriginal children of the province, like children everywhere, were able to amuse themselves with many playthings derived from plants. They liked to step on the bladders of Sea Wrack and other sea-weeds to make them pop, and to squirt water from the elongated sacs of *Halosaccion glandiforme,* another marine alga. Nuu-chah-nulth children dried the floats of Giant Kelp and threw them on the fire to make them explode "like firecrackers". Nuxalk girls played a game with Scarlet Paintbrush flowers: one girl held the flower while girls on the opposite team sang a song and tried to make her laugh or smile. Okanagan children played a "wishbone" game with little hooked stems of wild buckwheat. Children in many parts of the province made blow-guns from hollowed stems of elderberry or Cow Parsnip, using pieces of kelp or other vegetation as ammunition. (Caution: blowguns made from Cow Parsnip stems can be dangerous. See the warning on page 133.)

Adults, too, used many plants and plant products in their recreational activities. Okanagan men played a type of throwing game using a spoked wheel made of Rocky Mountain Juniper as a target. They also played a pin-and-ball game, with Red Hawthorn spines as pins and Tule-stem balls. On the coast, people played a throwing game using spears made of Salmonberry stems and targets of Bull Kelp bulbs. Young men of the Squamish, Ditidaht and other south-coastal nations competed in an endurance contest with Sword Fern fronds, pulling off the leaflets one at a time and saying "*pála*" ("one") with each one to see who could pull off the most in one breath. The Nuu-chah-nulth played a kind of hockey on the beach with "pucks" carved from sections of dried *Laminaria* kelp stems or hold fasts. Tree fungi were used as balls in a Kwakwaka'wakw handball game and as targets in a Sto:lo bow-and-arrow game.

The use of plants as biological indicators was common and widespread. Here are just a few examples: According to Squamish lore, when the Stinging Nettle shoots were just appearing above the ground, the seals were having their young. In Okanagan tradition, the blooming of the Mock-oranges signified that the marmots were fat and ready to be hunted. When the larches turned yellow in the fall, pregnant female bears were entering their dens for the winter. Finally, to some people, the shimmering of Trembling Aspen leaves when there was no perceptible wind was a sign of an impending storm.

Some technological uses of plants defy classification and can only be described as unique applications. Two examples are: the Secwepemc used living, growing White Clematis vines to straighten and strengthen implement handles; and the Okanagan people placed prickly-pear cactus around cache poles to keep rodents and larger mammals away from their food supplies.

Harvesting Plant Materials

First Peoples harvested most plant materials by hand or with the aid of simple tools. In the days before European contact, they felled trees by chopping around the base with adzes, or with stone mauls and chisels made of shell, horn or stone; some groups also used controlled burning to fell trees. People also harvested wind-toppled logs or they split boards and half-logs from standing trees using a series of graduated hardwood wedges. Iron was available, even before Europeans arrived, but it was very rare. Naturally, the introduction of larger quantities of iron for adzes, axes and other tools greatly facilitated the harvesting of timber, and the processing and use of plant materials in general.

Bark and bark-fibre were most easily gathered in late spring and early summer, when the running sap allowed the bark to separate easily from the underlying wood. People obtained sheets of bark, such as that of Paper Birch, by making two horizontal cuts some distance apart around the tree and one vertical cut between them, then peeling off the bark as a single sheet; they did this without harming the underlying inner bark or growing cambium tissues of the tree. They cut Bitter Cherry bark off the tree in sheets, as above, or in thin horizontal or spiral strips. Cedar bark (Yellow and Red) could be removed in two ways: for most uses, people made one cut and pulled the bark up and off the tree in a long, vertical strip that tapered until it broke off at the top; for roofing material, they cut it at the top and bottom and pulled off rectangular sections.

Aboriginal peoples usually collected fibrous stems and leaves in late summer and fall when the plants were fully mature. They cut, sorted and bundled the stems or leaves, then hung or spread them in the sun to dry. They dug up roots for basketry and cordage, using wooden or iron digging sticks similar to the type used for harvesting edible roots. They preferred long, straight roots of even thickness and with few side-branches. Trees widely spaced, growing in sandy or loamy soil, were said to have the best roots. People gathered roots in any season, but late spring was said to be an ideal time because the bark peels off best then. After harvesting, they heated the roots over coals to prevent the woody tissue from darkening, removed the bark and split the roots, and then dried them for later use.

People gathered most other plant materials fresh during the growing season. They replenished some materials frequently, as they were needed – such as materials used for bedding, floor coverings, steaming-pit linings and insect repellents. They harvested others in quantity – including some dyestuffs, tinder and cleansing agents – then dried and stored them until needed.

As a general rule, men harvested wood and built the larger structures, and they also collected the larger sheets of bark for canoes and roofing. The women were usually responsible for harvesting and preparing fibrous plant materials, such as inner cedar bark and various leaves, stems and roots for making baskets, mats and clothing. But if the fibres were to be used to make fishing lines and nets, or in some way involved with fishing, hunting or woodworking, the men might gather and process them as well. Other materials were gathered by men, women or children, depending on the effort involved and on who would be using the final product.

Important plant materials, such as cedar bark on the coast or Indian Hemp fibre in the interior, might be harvested during an outing of a day or several days by an entire family or a party of men, women and children in a village group. Such gathering expeditions were often combined with other activities – hunting, fishing, digging edible roots or picking berries. The materials could be processed in the field to the point where they would not spoil or deteriorate, then carried home to be worked on when the harvesting seasons were over and the weather was too poor for travelling.

Waiting for the Canoe by E.S. Curtis. Virginia Tom of Hesquiat Harbour and a woman from Opisaht dressed in traditional clothing made from Red-cedar bark.

The gathering and preparation of the materials could be difficult and time consuming, but was undoubtedly made more enjoyable when undertaken by a group of people and turned into a social event. The burden of the work was certainly eased with songs, such as one sung by Haida women when pulling off cedar bark, roughly translated as: "We want a long strip; go up high; go up high!"

Technologically important plants growing in the territory of a village group were considered to be the property of that group, and other people wishing to harvest them were obligated to ask permission from the villagers. It is likely that families and individuals could own patches or stands of plants that they used for materials, as they could patches of edible plants in some areas.

According to aboriginal religious traditions, plants, like animals, were believed to have "souls" and to be capable of thought and feeling just as people are today. Aboriginal people recognized the need to use and exploit natural objects, but they generally approached them with reverence and respect, and rarely used them wastefully or without due appreciation. This belief is evident in the following prayer by a Kwakwaka'wakw woman to a young cedar tree from which she is about to harvest the bark (from Boas 1909).

Words of Praise – Prayer to a Young Cedar

The woman goes into the woods to look for young cedar trees. As soon as she finds them, she picks out one that has no twists in the bark, and whose bark is not thick. She takes her hand adze and stands under the young cedar tree, and looking up to it, she prays, saying:

Look at me, friend,
I come to ask for your dress,
For you have come to take pity on us;
For there is nothing for which you cannot be used, . . .

For you are really willing to give us your dress,
I come to beg you for this,
Long-life maker,
For I am going to make a basket for lily roots out of you.

I pray, friend, do not feel angry
On account of what I am going to do to you;
And I beg you friend, to tell our friends about what I ask of you!

Take care friend!
Keep sickness away from me,
So that I may not be killed by sickness or in war,
O friend!

The Kwakwaka'wakw bark harvester was careful not to completely girdle a cedar tree, because this would kill the tree, and the nearby cedar trees would curse the person who did it. All present-day users of plant materials in British Columbia would do well to emulate this traditional respect for natural resources.

Preparing Plant Materials

The harvesting of plant materials was an arduous task, but in many cases it was only the beginning of a long, painstaking process leading ultimately to a beautifully finished product, whether it be a cedar-wood canoe, a Tule mat, or a birch-bark basket. Imagine making a sweater with wool that you first had to go out and shear from the sheep, then wash and card, dye, spin, design and knit. That would be approximately equivalent to the task of an aboriginal basket maker or woodworker. Even the tiniest basket might take a day or more to complete; weaving a large basket would take many weeks, and carving a canoe might take an entire winter.

As mentioned earlier, in aboriginal cultures men usually performed woodworking activities, while women were responsible for processing the various fibrous materials. Woodworking techniques involved the use of chisels, adzes and knives for carving, wedges and mauls for splitting, and drills for gauging thickness and for sewing and pegging seams. Fire was also an important tool: controlled burning was often used to hollow out canoes and large feast dishes; and hardwoods, such as "Ironwood" (Oceanspray), Saskatoon Berry, Mock-orange and crabapple, were often "baked" over hot coals to make them even harder. Steam generated from red-hot rocks and damp vegetation, such as seaweed, was used to mould and bend woods in making such items as canoes, boxes, snowshoes and fish hooks. With careful work, steamed or water-saturated wood could actually be bent to an angle of 90 degrees, as was done to make the sides of kerfed cedar-wood boxes. The final step in making most wooden objects was to smooth and polish them. Shark or dogfish skin was often used for this purpose, but when it was not available horsetail stems could be used.

Wooden objects, especially in the Northwest Coast Culture Area, were often decorated with precise, symmetrical, painted designs or relief carvings, usually of themes from traditional stories or of family or clan crests, many of animal origin. Northwest Coast artifacts – wooden articles and baskets – were frequently standardized in form and size. This was particularly true in the Kwakwaka'wakw culture, where every particular kind of feast dish, every berry-drying frame and every elderberry basket had virtually the same dimensions as every other. Finger-widths, hand-widths, arm-lengths, and pieces of string and rope were the usual measuring devices.

The preparation of fibres and fibrous materials required such tools as knives, beaters and, in the case of stem fibres, simple spindles. Some people used a type of suspended warp loom for weaving robes and blankets of cedar bark, and for weaving Mountain Goat wool and other animal products. For sewing Cattail and Tule mats, people used long hardwood needles and wooden "mat-creasers" to press the leaves or stems down around the needle, allowing it an easier passage and preventing the mat material from splitting. They used awls and bone needles to stitch birch-bark baskets and other items made of bark sheets. But certainly, the hands were the most important tools for working fibrous materials.

Sheets of bark were generally worked on as soon as they were harvested, before they became too brittle. Red-cedar bark sheets for roofing were flattened while still green by inserting sticks at intervals through the fibrous tissue, and weighting them with rocks until they dried. Birch bark and other barks used for canoes and baskets were cut and shaped according to the natural tendency of the bark to curl outward, so that the inner surface of the bark formed the outside of the vessel. To store these sheets of bark, people rolled them up against the direction of the natural curl; otherwise, the bark would roll up too tightly and, when dry, would not unravel without breaking. People stitched the seams of bark vessels with lengths of root and bound a light wooden frame to the rim to add strength and support. They designed special bark baskets by etching patterns into the bark surface, using fancy edge-stitching and overlaying Porcupine quills, cherry bark or other materials.

Most other fibrous materials, after being processed to a certain point, could be dried and stored. Dried fibres could be made flexible simply by soaking them in warm water for a short time. Aboriginal harvesters extracted stem fibres, such as those of Stinging Nettle and Indian Hemp, by splitting and scraping the stems, then carefully breaking away the brittle pithy inner tissue in short lengths, leaving the fibres to be cleaned and spun into string. They spun the fibre by rolling the strands on the bare thigh, sometimes using a long, hand-twirled spindle with a large hardwood or bone whorl as a flywheel to maintain an even tension. Lengths of fibre could be spliced together during the spinning process to produce a continuous strand. Individual strands could be twisted or plaited together to form a multi-ply twine or rope, suitable for fishing lines and nets, for sewing or twining mats and capes, or for binding implements.

Bark fibres and fibrous leaves, stems and roots were used whole, split into pieces of uniform width, or in the case of cedar bark, sometimes pounded into soft cottony strands. Aboriginal peoples applied a variety of weaving and sewing techniques to these materials, according to the type of plant they used, the type of product they were making and the cultural practices of their group. The Kwakwaka'wakw and Nuu-chah-nulth specialized in weaving robes of Yellow-cedar bark with a simple loom. They hung long strands of the pounded bark over a cord and twined them together at intervals with double strands of tightly twisted Red-cedar bark or Mountain Goat wool. The interior peoples used a similar technique to weave capes and clothing of such materials as Silverberry bark or White Clematis bark twined with Indian Hemp string. The Coast Salish people made robes and blankets chiefly of Mountain Goat wool, duck and goose down, and the hair of small, domesticated dogs, but they sometimes spun plant fibres, such as Fireweed and Cattail seed fluff, in with these materials. They used both twilled checker and twining weaves; in the latter, Red-cedar-bark cord was often used as the active weaving weft material.

Coastal peoples wove cedar-bark mats and sometimes baskets and hats in simple checkered and diagonal checkered styles. They often interspersed dyed bark strands with the undyed elements to produce geometric patterns, and also incorporated twilled designs, especially on the borders. The Salish peoples of both the coast and interior traditionally made Cattail and Tule mats. They laid the leaves or stems side by side and sewed them together with a long needle; with Tule, they sometimes twined the stems together at intervals with string. To keep the cut ends from tearing apart, weavers often folded them over and bound them. They often added plaited strips to the edges of mats.

Most Northwest Coast basket makers used twining techniques. They made some baskets with an open weave, to allow air circulation or the draining of liquids, and others with an extremely tight weave. They decorated their baskets by superimposing naturally or artificially coloured materials, such as Bear-grass and surf-grass and the stems of various true grasses, in intricate patterns over the basic weave. The Nuu-chah-nulth, Haida and Tlingit of Alaska were especially proficient in fine, close basket twining. They made flared waterproof hats by the same technique.

The plaited edge of a Tule mat (made in Saanich).

The Salish peoples of the province were specialists in coiled basketry, which involves sewing together spirals of flat or upright coils of fibrous materials, such as split Red-cedar roots, with flat strips of the same or a different material. They sewed the coils so closely and carefully that the underlying material was completely covered and only the evenly stitched outer layer was visible. Salish basket makers usually decorated their creations with imbricated designs. For these they often used strips of natural red and black-dyed Bitter Cherry bark and straw-coloured grass stalks, such as those of Reed Canary Grass. Some people used Engelmann Spruce and White Spruce roots for coiled baskets similar in style to cedar-root baskets.

Many plant materials required little or no preparation. People used fresh plants for bedding, or to cover floors, line steaming pits, make drying racks or berry baskets, collect herring spawn or whip Soapberries. Some processing involved such minor tasks as trimming large branches or removing the midribs of Skunk Cabbage leaves. Bull Kelp used for storing liquids and for water conduction could be employed fresh; or, for greater durability, it could be cured by alternately drying it in the sun and soaking it in fresh water, and sometimes by rubbing it with oil. Materials for infant diapers, wound dressings and tinder were air dried before use and, in some cases, softened by pounding. Horsetails could be used fresh or dried for smoothing and finishing; and tree pitch was used fresh as a glue. Most of the plants used as insect repellents were burned while still green as a smudge or boiled to make a washing solution. Soaps and cleansing agents were used fresh, dried or in solution, as were the various plants used for their scents; aromatic plants were also burned as incense. Plants for animal poisons were usually steeped in water to make a solution. Most dyes and tanning agents were prepared by boiling the colouring material in water, sometimes with a urine mordant. Paints were usually made by pounding the dried plant matter to a powder and mixing the powder with ochre, fat or resin. Stains were made by macerating fresh materials, such as berries, flowers or leaves, and rubbing the moist pulp onto the object to be coloured.

In making almost any implement, container or article of clothing, people used a combination of different plant materials, all of which had to be gathered and processed. For example, a bow maker had to select, carve and mould the appropriate wood, and prepare a strong bowstring, perhaps of Indian Hemp fibre. He might bind the haft with Bitter Cherry bark, which would be glued in place with tree pitch and tied with more fibre. Finally, he might polish the wood with horsetail stems and decorate it with a vegetable paint or stain. A Nuu-chah-nulth basket maker might use the inner bark of Red-cedar as a foundation for her product, one type of sedge for the bottom, another for the sides and top, Bear-grass for the edging, and the same materials dyed, or other materials of different natural colours, for weaving in designs. The potential for variation was limitless.

Although basketry is not as prevalent an art as it once was, there are still many talented basket makers in communities throughout the province. Basket making, like woodworking, is an important part of peoples' cultural heritage, and will certainly endure.

Trading Plant Materials

Every aboriginal group in the province had access to a great variety of plant materials, but of course, they valued some more than others. In areas where prized plants did not occur naturally, people often obtained them by trade from neighbouring groups, either in the form of raw materials or as finished products. Often, too, the people of one group would be particularly skilled in constructing a certain type of product and would be able to trade it to neighbouring groups even when the raw materials were just as readily available to the neighbours.

The exchange of plant materials and other economic products took place at all levels – in family and village groups, between villages in the same language group, among the different language divisions on the coast and in the interior, and even between coastal and interior groups. Some groups, especially those in the transitional zone between the coast and the interior, acted as middlemen, buying the products of one neighbouring group and reselling them to another. With the coming of Europeans and the accompanying influx of new trade goods and improved transportation routes, the exchange of plant products became even more widespread.

Examples of the trading of raw materials are numerous. The peoples of the Olympic Peninsula and other parts of Washington gathered Bear-grass, used for fine trimming and imbricating baskets, dried and tied it in bundles, and traded it to the Coast Salish and Nuu-chah-nulth of Vancouver Island, who prized it greatly. The Lower Stl'atl'imx transported dyed and natural Red-cedar bark, and Western Yew, Vine Maple and Yellow-cedar wood from the coast and traded them to the Upper Stl'atl'imx in exchange for Indian Hemp fibre, Silverberry bark and certain grasses used in basketry. These they then sold to the Squamish, Sechelt and Comox peoples of the coast. The Upper

Stl'atl'imx sometimes sold Yew wood and other coastal materials, as well as Mock-orange wood and some of the other products within their own territory to their Secwepemc neighbours. In Secwepemc territory, people exchanged basketry materials, such as Red-cedar roots, for salmon and other products at large tribal gatherings, including one held annually at Green Lake, as reported by James Teit (1909). The Haida obtained Rocky Mountain Maple wood and Cottonwood from the mainland Tsimshian-speaking peoples, along with Eulachon grease and other animal products. In turn, the Tsimshian acquired large Red-cedar bark sheets for roofing and other materials and foods from the Haida.

The trading of finished products was equally common and widespread. The Lower Stl'atl'imx, for example, sold split cedar-root baskets and birch-bark baskets from the interior to their coastal neighbours. The Haida, renowned for their wood-carving ability, sold superbly crafted kerfed cedar oil-boxes and large canoes to the Tsimshian. Similarly, the Nuu-chah-nulth sold their excellent canoes far and wide among their Salish neighbours to the south and even to the Chinook on the lower Columbia River and to the peoples along the central Oregon coast. The Nuu-chah-nulth and the Kwakwaka'wakw traded their famous Yellow-cedar-bark robes both north and south along the coast. The Secwepemc sold birch-bark baskets to the Okanagan and bought Indian Hemp bags, baskets and mats from them.

Not only were finished products exchanged among aboriginal groups, but so were the techniques and skills involved in making them. For example, the Nuxalk learned from the Carrier and Tsilhqot'in how to make birch-bark baskets and canoes. Some ethnologists suggest that the Coast Salish learned the technique of making twined baskets from the Kwakwaka'wakw and Nuu-chah-nulth; they believe that the Salish on the coast originally made only coiled baskets like their Interior Salish relatives. Similarly, the Kwakwaka'wakw and Nuu-chah-nulth may have learned the art of making Cattail and Tule mats only recently from their Salish neighbours. In many cases it is impossible to determine which group originated a particular skill or technique. Learning more about the development of botanical technologies among the First Peoples of our province could tell us much about the origins and inter-relationships of the people themselves.

ALGAE
Including Seaweeds,
and Freshwater and Terrestrial Types

One type of seaweed, Bull Kelp, was of major significance in the technology of coastal First Peoples; it is described in detail below. A number of other seaweeds were used in minor capacities. Nuu-chah-nulth children dried the floats of Giant Kelp and put them in the fire to explode; they sometimes called them "Indian firecrackers". Haida children liked to squirt water out of the sacs of Sac Seaweed, pretending (at least within the last century) that they were the nipples of a cow. Children also stepped on plants of brown algae and the swollen bladders of Sea Wrack (also called "Rockweed") to make them pop. Some Nuu-chah-nulth peoples played "beach hockey" along the shore, using pucks of dried sections of the stems, or small, hard balls carved from the holdfasts of brown algae, including *Laminaria* species.

Most coastal First Peoples piled Sea Wrack and other seaweeds, such as Red Laver, over hot rocks to generate steam for cooking, bending and moulding wood, or medicinal sweat-baths. They used Giant Kelp to collect herring eggs during the spawning season, eating the fronds with the eggs; they also used Boa Kelp, Sea Wrack and Bull Kelp to collect herring eggs. Seaweeds were ideal for covering fish in boats or

Sea Wrack.

Several seaweeds.

"Green pond slime".

canoes to keep them cool on hot days. The Squamish rubbed Sea Wrack on their fishing lines to remove the human scent. When aboriginal people began tending vegetable gardens following contact with Europeans, some fertilized their potatoes and other vegetables with seaweeds such as Boa Kelp.

Several interior aboriginal groups, including the Nlaka'pamux, Carrier and Tahltan, obtained a green dye for basket materials by boiling rotten, decayed wood, presumably extracting terrestrial green algae from the wood. The Nlaka'pamux also used "green pond slime", from colonies of *Spirogyra* and other aquatic algae, as a green pigment.

Bull Kelp
(Brown Algae)

Nereocystis luetkeana
(Phaeophyceae)

Botanical Description

Bull Kelp is one of the largest marine algae. The slender, unbranched stalk grows up to 30 metres long, and is attached to the bottom by a stout, rootlike holdfast. The stalk is solid at its base, becoming hollow and increasing in diameter towards the upper end. It terminates in a large, spherical float, up to 15 cm in diameter. Attached to the upper surface of the float are two clusters of flat, elongated leaflike blades,

each 6 to 15 cm broad and up to 4.5 metres long. The entire plant is golden to dark brown.

Habitat: rocks in the upper subtidal zone and to a depth of 20 metres or more, often growing in thick beds of several hectares in sheltered inlets and bays.

Distribution in British Columbia: along the entire Pacific coast.

Aboriginal Use

Virtually all coastal groups in the province made fishing lines, nets, ropes and harpoon lines from Bull Kelp. They dried and cured the long, ropelike stalks, then spliced or plaited them together. Curing methods varied: some Coast Salish peoples alternately soaked the kelp in fresh water and dried it over a smoking fire; the Nuu-chah-nulth dried the kelp and soaked it in dogfish or whale oil. Kelp lines were dried for storage, but had to be soaked in water before use, or they were too brittle. After soaking they became strong and flexible once again.

Haida fishermen used to keep a number of kelp lines of different lengths – one of about 350 metres, two of 200 metres and one or two of 20 metres. To catch deep-water fish, such as Sablefish (Black Cod), about ten people would tie their longest lines together and fasten all the hooks near the end. Kelp was also used to make anchor lines, and fishermen would often fasten their boats to living kelp plants firmly anchored on the bottom.

People all along the coast used the hollow upper stalks and bulblike floats of Bull Kelp as storage containers for Eulachon grease, fish oil and water, and after European contact, even molasses. They cured the stalks and floats by soaking them in fresh water, then drying them. They poured the liquid into the float, or into a length of stipe with one end tied, and sealed it by tying the open end. The filled tubular containers could be coiled and laid in wooden chests or hung in a convenient place. To obtain some of the contents, one simply untied the knot and squeezed out the required amount of liquid, then retied the knot for further storage.

Some Nuu-chah-nulth groups used fresh kelp floats as moulds for deer suet. They poured the melted fat in through the hollow stem, allowed it to harden, then broke the kelp away, leaving a bulb of suet ready for storage. People sometimes made a salve of Cottonwood buds boiled in deer fat. They poured the mixture in a

Bull Kelp fishing line, 400-500 metres long. The lengths of kelp are tied together with fisherman's knots and whipped with spruce root. C.F. Newcombe collected this from Haida Gwaii in 1914. (RBCM 1461)

kelp bulb and let it harden: the result was a fragrant ointment for protecting the skin from sunburn and windburn.

Some Coast Salish groups used kelp to form the warp in baskets, mats and blankets, but its brittleness when dry made it generally unsuitable for this purpose. The Coast Salish, Ditidaht and other groups made halibut hooks by placing split Douglas-fir or Western Hemlock knots (dense branch ends extracted from rotten logs) inside hollow kelp stalks and burying the pieces overnight in hot ashes to make them easy to mould into hook forms. In a similar manner, the Straits Salish placed the ends of Yew-wood bows inside lengths of kelp and steamed them to mould and bend to the desired shape. Squamish fishermen used kelp blades to keep fish fresh and moist in the canoe. The Comox used them to line steaming pits to flavour the food and help generate steam.

The Kwakwaka'wakw sometimes used the hollow tubes in their steaming pits to add water to the hot rocks at the bottom during the cooking process. They also buried the stalks under the dance house floor to achieve special effects during winter dances, such as having voices coming from the middle of the fire. The Nuxalk used lengths of kelp as water conduits; their name for the modern garden hose means "kelp".

Kelp was also employed in various recreational activities. The Straits Salish and the Haida used the stalks as targets in throwing games. Coast Salish and Kwakwaka'wakw children used them as ammunition for their toy elderberry-stem blowguns; they hung the stalks around their necks and tore off small pieces as they needed them.

LICHENS
(Lichenes)

Wolf Lichen was widely used by British Columbia First Peoples as a dye and pigment, so it is discussed in detail below. Other lichens were also used as materials, but on a more restricted basis. Black Tree Lichen and its close relative, the light green Old Man's Beard were sometimes used by the Stl'atl'imx and other Interior Salish peoples for weaving items of clothing, such as ponchos and footwear; but they were not considered a high-quality material, and were usually used only by poorer people who could not obtain skins for clothing. Lichens were usually interwoven with some stronger fibrous material such as Silverberry bark. The Secwepemc and Nuxalk used these species as false whiskers and artificial hair for decorating dance masks and, especially by children, for masquerading. The Sechelt used Old Man's Beard for baby diapers, and put it on the fire when they desired smoke. The Haida used it and a similar lichen, *Usnea longissima* (also called Old Man's Beard), to strain hot pitch to remove impurities before it was used as a medicine; and the Haida used this lichen as bedding when camping.

According to one source, some Coast Salish peoples used a species of *Usnea* to make a dark-green dye, and Black Tree Lichen mixed with Wolf Lichen to make a yellow dye, but this use has not been substantiated by contemporary consultants. The Haisla and other coastal peoples reportedly made pigments from certain black and yellow lichens that grow on rocks and trees. They mixed the

Black Tree Lichen. (See also the photograph on page 83.)

powdered lichen with salmon eggs and used the resulting paint on spoons, bowls and totem poles. People used some species of *Usnea* and *Alectoria* to wipe the slime off fish and for protecting food in earth ovens.

Wolf Lichen
(Parmelia Family)

Letharia vulpina
(Parmeliaceae)

Other Names: "Wolf Moss", Common Wolf Lichen.

Botanical Description

Wolf Lichen is short, upright, densely branching and bright yellowish-green. It grows 2.5 to 5 cm tall on coniferous trees and wood. The individual branches are wiry and round in cross-section, bearing greenish powdery fragments called soredia. The fruiting discs, when present, are small (about 5 mm across), brown and flat. This species contains a

brightly coloured but poisonous lichen acid known as vulpinic acid, which is responsible for its striking colour. A similar species, Brown-eyed Wolf Lichen, occurs in British Columbia's interior and was probably used for dye as well. It commonly has spore-bearing discs and lacks the powdery fragments (soredia) of Wolf Lichen.

Habitat: branches, dead wood and bark of coniferous trees such as Douglas-fir, Ponderosa Pine and Western Larch; commonly grows with the long, hairlike Black Tree Lichen.

Distribution in British Columbia: throughout the interior, especially in montane forests and the dry areas of the south; it also occurs sporadically west of the Cascade Mountains, where it is generally stunted and less dense.

Aboriginal Use

Lichens are well known in many areas of the world as sources of dyes. Many lichen acids are either themselves brightly coloured or yield pigments when combined with certain chemical substances such as alum, chrome and tannic acid. The only lichen documented as being used to any extent for dyeing by the First Peoples of British Columbia was Wolf Lichen, which yields a brilliant yellow dye (although Brown-eyed Wolf Lichen may have also been used). Wolf Lichen was used by virtually all the interior peoples – the Ktunaxa, and the Salishan and Athapaskan peoples – as well as the Flathead Salish of Montana and the Blackfoot of Alberta. Some coastal groups also used it, when available. The Alaskan Tlingit obtained it by trade from the interior to colour their spruce-root baskets and Chilkat blankets. The Nuxalk obtained it by trade from the Ulkatcho Carrier, and in turn distributed it to their coastal neighbours.

The simplest way to extract the pigment is to boil Wolf Lichen in water, then steep the item to be coloured in the solution. The Okanagan sometimes added Oregon Grape bark to the water to intensify the colour. People used the pigment to colour basket materials, fur, moccasins, feathers, Porcupine quills, wood, and in modern times, cloth and horsehair for braided bridles. The Nlaka'pamux and perhaps some other groups used it as a face and body paint. They simply dipped the lichen in water and brushed it on the skin, or wet the skin and applied it dry.

FUNGI

Many First Peoples used tree fungi in their technology, especially Indian Paint Fungus, but they used mushrooms and fleshy fungi in only minor ways. According to one consultant, the Nlaka'pamux people sprinkled water on Inky Cap mushrooms to kill flies and other insects. Blackfoot men of Alberta once wore necklaces of golf-ball-sized puffballs, said to be prized for their delicate odour; these puffballs also inspired the designs on some Blackfoot tipis.

Indian Paint Fungus (Polypore Family)
Echinodontium tinctorium (Polyporaceae)

Botanical Description
A close relative of the bracket fungi, Indian Paint Fungus grows mainly on coniferous trees. Its fruiting structures are woody and rounded, averaging about 5 cm across, but sometimes much larger. It produces

spores in crowded circular pores on the undersurface. The external and internal tissue is bright red-orange.

Habitat: mainly on the trunks of living coniferous trees, such as Western Hemlock, Douglas-fir and true firs (*Abies* spp.).

Distribution in British Columbia: throughout the province and locally abundant in certain areas, such as in the Coastal Western Hemlock Zone .

Aboriginal Use

Indian Paint Fungus was used as a red pigment in many areas of the province, although in most cases its identity was established only from the descriptions given by elders or from published sources. Records of its use extend to the following groups: Okanagan, Secwepemc, Nlaka'pamux, Lower Stl'atl'imx, Straits Salish, Sechelt, Kwakwaka'wakw, Nuu-chah-nulth, Oweekeno, Haida, Tsimshian and Tahltan. Undoubtedly, other peoples used it as well. The Haida reportedly obtained it by trade from the Tsimshian on the mainland. The main use of the pigment was as a face paint, for cosmetic purposes, and to protect the skin from sunburn or insect bites. The Tahltan used it to absorb some of the sun's glare off the snow, reducing the chances of snow blindness.

The standard method of preparing the paint was to dry the fungus, usually by heating it in a fire, then powder it and mix it with melted fat, grease or pitch. The resulting salve was smeared directly on the face. The Tahltan simply applied a thin layer of suet to the skin, then dabbed powdered fungus over the top. The Saanich mixed Red-cedar bark and Red Alder bark with the fungus before heating it and used the powder for tattooing. The Kwakwaka'wakw mixed it with Western Hemlock gum as a face paint, but also made a general-purpose paint by placing the fungus on hot stones, covering it with wet Lady Fern fronds and leaving it until it became a red powder. Like most natural dyes and pigments, Indian Paint Fungus is seldom used today.

Bracket Fungi

(Polypore Family)

Fomitopsis, Ganoderma, Polyporus and related species
(Polyporaceae)

Other Names: shelf fungi, polypores.

Botanical Description

Bracket fungi are found on living or dead tree trunks, as well as on stumps and fallen timber. The underlying portion, the mycelium, penetrates deeply into the wood, softening it and ultimately causing it to rot and crumble. It can often be seen as a white or yellow cottony substance beneath the bark or in the cracks of rotting wood. The fruiting portion is more conspicuous, consisting of woody or leathery flat to hoof-shaped "brackets". The upper surface is hard, varying from light brown to orange-brown to greyish, depending on the species. The lower surface is usually white, with innumerable closely packed, rounded pores in which the spores are produced. Many species are perennial – the size of the fruiting structures increases by annual additions to the pore surface and outer edge. Some bracket fungi attain a size of 30 cm or more across. They can usually be removed from the tree or log by applying pressure to the upper surface.

One species of bracket fungi: *Ganderma applanatum.*

Habitat: living and dead wood of both coniferous and deciduous trees; many types show a preference for coniferous or deciduous hosts, and some for individual species of trees.

Distribution in British Columbia: throughout the province, especially in moist woods.

Aboriginal Use

Ethnographers and contemporary aboriginal people seldom distinguish the different species of bracket fungi other than by specifying their host tree. A number of aboriginal groups used the corky inner tissue of some species as punk for "slow matches". When ignited, the tissue smoulders for many hours. One can kindle a fire simply by blowing on the smouldering fungus. The Kwakwaka'wakw used a fungus growing on Grand Fir, while the Lower Stl'atl'imx used a soft fungus, "as yellow as sulphur", possibly Cinder Conk, which was used as tinder and a "slow match" by Gitxsan and Secwepemc peoples, among others (see the photograph on page 156; see also Gottesfeld 1992). Another species used similarly by First Peoples of British Columbia is Tinderwood Polypore, which has been employed commercially in manufacturing German tinder or punk sticks for lighting cigars and pipes on windy days and for touching off fireworks. Aboriginal people used clam shells, cedar bark or rolls of birch bark to contain the fungus punk for transport. The Blackfoot of Alberta carried their tinder fungus in a covered buffalo horn lined with moist, rotten wood.

The Nlaka'pamux reportedly did not use fungus as tinder, although they had heard of foreigners using it. But they did use a type of fungus from Cottonwood trees, possibly *Ganoderma* species, to tan buckskin. If the Cottonwood fungus was not available, any type of bracket fungus would do. Annie York described the process:

They chopped the fungus and mixed it with deer brains and water. Then they soaked the hide, already stripped of hair, in this mixture for a few hours. Adding a small amount of fish oil or, in modern times, butter, they left the skin to soak for four days. On the fifth day, the tanners removed the hide, laced it to a frame and worked it with the hands – first one side, then the other – until it was dry. Then they repeated the entire process. After the second soaking, they smudged the hide over a fire made with bracket fungus and rotten wood, covered with earth to prevent the hide from burning. When it had been smoked on both sides, they soaked the hide again in the fungus/brains/water mixture, and again stretched and worked it with the hands. By this time the hide was soft and white, suitable for making clothing or bags, although a perfectionist might subject it to the entire process of smoking, soaking and stretching one more time before using it.

The Sechelt tanned hides by burying them for several weeks with a small, scalloplike polypore, Turkey Tails. Other groups may have also used bracket fungi for tanning. Some people also burned tree fungi as a smudge against mosquitoes.

The Haida used the felt-like mycelium of a fungus growing on rotten Sitka Spruce – they rubbed the mycelium into a soft paste to make caulking for canoes and oil boxes. The Squamish used the corky inner tissue of various bracket fungi for scrubbing their hands. Nuxalk people painted faces on large specimens of bracket fungi, attached miniature bodies of cedar bark to them, and used them as dance symbols in a special "fungus dance" of the Kusiut ceremonials. Other coastal peoples also used fungi as ceremonial objects. The Tlingit used the large, woody spore-bearing bodies of *Fomitopsis officinalis* to carve shaman's grave guardians (see Blanchette et al. 1992).

The Sto:lo, Haisla and many other peoples used bracket fungi as targets in a variety of spear-throwing and ball games. Many coastal peoples attributed protective properties to these fungi, believing that they could deflect evil thoughts and that they caused echoes in the woods.

MOSSES
(Bryophyta)

With the exception of the distinctive and highly absorbent sphagnum mosses (discussed in detail on the following pages), most ethnographic sources and aboriginal people do not distinguish between the many different species of mosses and liverworts. Mosses were used throughout the province for numerous general household tasks, such as lining steaming pits and generating steam for cooking and moulding wood, wiping the slime off fish, covering floors, stuffing mattresses and pillows, and lining baby bags and cradles; they were also mixed with pitch to caulk canoes. Within the last two centuries, aboriginal people mixed mosses with mud to chink log cabins. Groups with easy access to sphagnum moss usually preferred it for these jobs, but if it was not available, they used any kind of moss, with preference to bulky, absorbent types. Diamond Jenness (n.d.) reports that the Saanich people used Menzies' Tree Moss to make a yellow dye for basket materials, and G.T. Emmons (1911) records that the Tahltan used a *Hypnum* species to line the cradle of a new-born baby, although James Teit (1906b) states that they used sphagnum for this purpose. The Flathead Salish of Montana used Rough Moss to line cradle boards and padding inside baby diapers; it was said to last 12 hours without needing changing.

Sphagnum Mosses
(Sphagnum Family)

Sphagnum species
(Sphagnaceae)

Other Names: "Indian sponge", "baby moss", "diaper moss", peat moss, bog moss.

Botanical Description

Sphagnum mosses are large, attractive mosses of acid bogs, fens and muskegs. Their long, weak stems are crowded together so that the plants form soft cushions and extensive beds, often covering several hectares. Alternate clusters of branches are borne along the stem; at the top is a dense rosette of branches. The plants vary in colour from whitish to golden to yellow-green to deep reddish purple, depending on the species and environmental conditions. The leaves are small, veinless and only one cell-layer thick, but with a unique structure readily visible under a low-power microscope. They consist of two kinds of cells: one is large, colourless and hollow, with pores on the surfaces that readily allow air and water to pass through; and the other is small and elongated, containing active cell contents and chlorophyll, and forming a network around the large empty cells. It is the large cells that give sphagnum mosses their amazing absorbent properties, enabling them to take up and hold water like a sponge. The spore cases are dark brown and globular, on thick stalks. There are many species of sphagnum in the province: Brown-stemmed Bog Moss, Small Red Peat Moss and Spread-leaved Peat Moss are among the most common.

Habitat: shaded to open swampy or boggy sites in acid soil from sea level to subalpine elevations.

Distribution in British Columbia: in suitable habitats throughout the province.

Aboriginal Use

The soft, absorbent qualities of sphagnum mosses made them ideal for use in personal hygiene and baby care. Virtually all First Peoples preferred sphagnum over other types of mosses for bedding, sanitary napkins, wound dressings and baby diapers. The northern groups, such as the Tahltan, especially used them. They had access to large muskeg areas where sphagnum is a dominant ground cover. Tahltan women, about to be confined for childbearing, gathered and stored large quantities of sphagnum. They used it to carpet the lodge in which the baby was to be born, to wipe the newborn baby's skin and to line the bark cradle. Mothers used spagnum mosses as diapers, placing a bunch between their babies' legs and holdin it there with a soft strip of animal skin, such as a marmot's. This diaper would last for many hours. If necessary, it could be washed, dried, and re-used. The Carrier and others used only the light-coloured sphagnums for diapers; they said that the red ones caused bad sores (see Gottesfeld and Vitt 1996).

Small Red Peat Moss.

FERNS AND THEIR RELATIVES
(Pteridophyta)

Almost all aboriginal peoples in British Columbia used horsetails, Sword Fern and Bracken Fern in their technology, so these plants are discussed in detail on the following pages. People also used several other ferns and their relatives on a more limited basis.

In modern times, the Nuxalk and Nuu-chah-nulth made wreaths and Christmas decorations from Running Clubmoss, although Nlaka'pamux elder Annie York said that keeping it in the house would bring bad luck. Other people sometimes used this plant for necklaces and belts.

The Nlaka'pamux, as well as the Haida, the Tlingit of Alaska, and the Makah and Quinault of Washington, imbricated baskets with the shiny black stems of Maidenhair Fern.

Haida Sitka Spruce-root basket imbricated with Maidenhair Fern stems in a whale design. Made by Isabel Edenshaw (Florence Davidson's mother). (RBCM 18359)

George Dawson (1891) records that the Secwepemc used Lady Fern to make a black dye, although he gives no details as to how they prepared it. The Kwakwaka'wakw placed Lady Fern fronds over Indian Paint Fungus when heating it to make a red paint.

The Kwakwaka'wakw and Coast Salish used the thin, wiry roots of the Spiny Wood Fern as tinder for making a "slow match" (see also the account for bracket fungi). A bundle of the dried roots was ignited and enclosed in a mussel or clam shell. The shell could then be buried or carried on trips and the fern roots would smoulder for several days.

Berry pickers sometimes used fern fronds to cover and press down the berries in their baskets to prevent them from spilling. Fern fronds, such as those of Spiny Wood Fern, were also used to pack around freshly caught fish to keep them moist and cool. To hunters and travellers in Okanagan country, the presence of ferns of any type indicated water nearby.

Common Horsetail	***Equisetum arvense***
Scouring-rush	***E. hyemale***
Giant Horsetail	***E. telmateia***
(Horsetail Family)	**(Equisetaceae)**

Other Names: "Indian sandpaper" (all); Field Horsetail (*E. arvense*).

Botanical Description
Horsetails are herbaceous perennials with deep, spreading, dark-coloured rhizomes and conspicuously jointed stems. Depending on the species, the stems are either all alike or of two kinds: one fertile and lacking chlorophyll, and therefore whitish in colour; and the other vegetative. The stems are usually erect, sometimes branching, hollow (except at the nodes) and cylindrical, with many regular longitudinal grooves. The walls of the outermost layer of cells are impregnated with silicon, and often there are well-developed silicified ridges between the grooves, making the plants rough to the touch. The leaves are small and scalelike, growing in whorls at each node and fused at the base to form a notched sheath, usually with dark markings. When branches are present they are in whorls, borne at the nodes. The spore cases are borne in terminal conelike structures, on separate fertile shoots or at the tips of the vegetative growth. Common and Giant horsetails are branching species with separate fertile and vegetative shoots. Giant Horsetail is the larger, its vegetative stems growing to more than 2 metres tall; those of Common Horsetail can grow to 70 cm tall. Scouring-rush is a non-branching species up to 1.5 metres tall, with stems all of one type.

Common Horsetail.

Scouring-rush stalks.   Giant Horsetail.

Habitat: Giant Horsetail and Scouring-rush grow in low, wet ground in swamps and along streams and seepage areas, while Common Horsetail grows in a wide variety of habitats, often along roadsides or in wasteland areas.

Distribution in British Columbia: Giant Horsetail is widely distributed along the coast as far north as Haida Gwaii, but does not occur in the interior; Common Horsetail and Scouring-rush are widespread throughout the province.

Aboriginal Use
The rough, silicon-impregnated cell walls of horsetails make them ideal as abrasives for smoothing and polishing surfaces. Most aboriginal people today equate them with sandpaper. Coastal peoples and some interior groups used horsetails to polish wooden objects – such as canoes, feast dishes, boxes, spoons, arrow shafts and points, gambling sticks, and within the last century, knitting needles – although they preferred using dogfish (shark) skin. Coastal groups used all three species, especially Giant Horsetail. Interior peoples used Common Horsetail and Scouring-rush; they did not have access to Giant Horsetail.

Interior Salish peoples used horsetails for polishing and sharpening bone tools and for smoothing and finishing soapstone pipes. For the latter, the Nlaka'pamux sometimes used a mixture of grease and Lodgepole Pine pitch along with the horsetails. The Okanagan coated their pipes with salmon slime, allowed it to harden, then rubbed the surface with the horsetails. The Okanagan used horsetails to polish their fingernails. The Stl'atl'imx used them to sharpen arrowheads.

Some groups also used the black underground rhizomes of horsetails in the decorative imbrication of baskets. The Tlingit of Alaska used Marsh Horsetail rhizomes and probably those of other species to imbricate their fine spruce-root baskets; and the Dena'ina of Alaska used horsetail stems to decorate their birch-bark baskets. The Coast Salish peoples of British Columbia and Washington used the rhizomes and stems of Giant Horsetail for their baskets, and the Sanpoil-Nespelem, a Washington Okanagan group, used Scouring-rush rhizomes to imbricate both baskets and storage bags. The Blackfoot of Alberta crushed the stems to make a light pink dye for colouring Porcupine quills. The Tahltan, and probably other peoples, made whistles out of the stiff hollow stalks of Scouring-rush.

Sword Fern
(Fern Family)

Polystichum munitum
(Polypodiaceae)

Botanical Description
Sword Fern is a perennial with stout, fleshy rhizomes. The fronds grow in clumps; they are evergreen, coarse and erect, often 60 cm or more in length. The stems are greenish and scaly. The pinnae are numerous, alternate, toothed and attached to the stem at a single point. Each has a prominent projection or "hilt" at the base of the upper edge. The sori are numerous, usually occurring in two rows on the undersurface of the upper pinnae, and covered with membranous, umbrella-like structures called indusia. The general shape of the frond is long, flat and tapering to a point at the top.

Habitat: common in damp, rich woods and on shaded slopes, but also occurs as a smaller form on open rocky exposures; generally confined to lowland forests.

Distribution in British Columbia: widespread and extremely common west of the coastal mountains; a dominant fern in lowland coastal forests rich in humus.

Aboriginal Use

The long, stiff fronds of Sword Fern were gathered in quantity by most coastal groups and by some interior groups – such as Stl'atl'imx and Nlaka'pamux – within the range of the species. They used the fronds to line steaming pits, storage boxes, baskets and berry-drying racks, to lay fish on and wipe them, to cover the floors of summer houses and dance houses, and to sleep on. For bedding, people sometimes wove the fronds into crude mats. The Nlaka'pamux often copied the Sword Fern pattern in the designs on their split cedar-root coiled baskets.

Squamish children and those of other southern coastal groups played an endurance game with the fronds to see who could pull off the most pinnae, saying "*pála*" with each one, in a single breath. The Squamish sometimes call this fern *pála-pála*. For the Ditidaht and some Coast Salish peoples, the *pála* contest was more serious: it was part of a young man's training, teaching him to hold his breath for a long time. This skill prepared young men for diving down to the base of a Bull Kelp to cut the stipe off the ocean floor.

Bracken Fern.

Bracken Fern
(Fern Family)

Pteridium aquilinum
(Polypodiaceae)

Other Name: Brake Fern.

Botanical Description
Bracken Fern is the largest, most common fern in British Columbia, often growing more than a metre and a half tall. The thick, fleshy rhizomes are perennial, often 20 cm deep, running horizontally for long distances, frequently branching. They are black outside, with a white inner tissue and tough longitudinal fibres in the centre. The fronds grow individually along the rhizomes, and they have tall, smooth, light-green stems and coarsely branching pinnae. The fronds and lower pinnae are broadly triangular in shape. The pinnules are numerous and deeply toothed, and the sori, when present, are marginal.

Habitat: common in open forests and clearings at low and moderate elevations.

Distribution in British Columbia: generally throughout the province, except at very high elevations.

Aboriginal Use
The large fronds of Bracken Fern, like those of Sword Fern, were used for many household purposes. Aboriginal peoples throughout the province, particularly the Okanagan, used them to line steaming pits; they also used them to cover berry baskets, store dried foods on, wipe fish and make bedding. In the old days, the Ktunaxa made sunshades from Bracken Fern fronds. Within the last century, the Nuxalk used the dead fronds to mulch their potato hills. The Nuu-chah-nulth, Kwakwaka'wakw and Oweekeno saved the fibrous remnants from the edible rhizomes, dried them, and used them for tinder and as punk for "slow matches". Contained in clam shells or tightly bound in cedar bark, the dried fibres would hold a fire for many hours or even days. According to Diamond Jenness's notes (n.d.), the Straits Salish and Halkomelem used bundles of the fibres of Bracken Fern rhizomes for torches in early spring; at other times of the year they used bundled cedar sticks. Haida basket weavers used Bracken Fern fronds as the basis of a basket-design pattern.

CONIFERS AND THEIR RELATIVES
(Gymnospermae)

Yellow-cedar

(Cypress Family)

Chamaecyparis nootkatensis

(**Cupressaceae**)

Other Names: Yellow Cypress, Alaska Cedar, Alaska Cypress, Sitka Cedar, Sitka Cypress, Nootka Cedar, Nootka Cypress.

Botanical Description
Yellow-cedar is a large tree, usually 20 to 40 metres tall and 90 cm or more in diameter, although gnarled and stunted individuals occur in unfavourable habitats. The bark is thin and greyish-brown, tending to

shed in long, narrow shaggy strips. The wood is yellowish and pungent smelling. The branches are horizontal and well-spaced. The branchlets are flat and hang down, giving the tree a drooping appearance; they are covered with small scale-like bluish-green leaves that make them prickly to the touch. The cones are spherical, about 12 mm in diameter, with up to six thick, woody scales; the cones grow alone or in small groups. The seeds are about 3 mm long, golden-brown and winged. A Yellow-cedar tree looks similar to a Red-cedar, but it droops more, is shaggier looking and has a more pungent scent.

Habitat: damp coastal forests, usually at subalpine elevations; also in peat bogs and muskegs, and in rock crevices, where it is usually stunted.

Distribution in British Columbia: common in coastal subalpine forests from Vancouver Island to Alaska, mostly west of the Cascade and Coast mountains between 600 and 900 metres elevation, but extending to near sea level in the north, and reported from Slocan Lake in the Kootenays.

Aboriginal Use

Virtually all coastal First Peoples carved implements from the Yellow-cedar's tough, straight-grained wood. Yellow-cedar bows, especially from the roots, were popular trading items. The Fraser River Stl'atl'imx obtained them, or the wood to make them, from the Lower Stl'atl'imx of the Pemberton area; the Saanich got them from mainland Salish groups. The Sechelt made bows only from the wood of young Yellow-cedars. The Vancouver Island Salish carved paddles and, in modern times, knitting needles from Yellow-cedar wood. The Nuu-chah-nulth carved masks from it. The Kwakwaka'wakw used it for paddles, chests, dishes, fishnet hoops and totem poles. The Haisla and their neighbours used it to make paddles, pegs for Red-cedar boxes, masks and, recently, boat ribs. Some people made dugout canoes from Yellow-cedar logs, though almost everyone preferred Red-cedar. And the Haida made Yellow-cedar digging sticks, adze handles, paddles, dishes and, recently, bedposts.

The inner bark of Yellow-cedar has the same fibrous qualities as that of Red-cedar, but it is considered even more valuable because it is finer, softer and lighter in colour when dry. It was pulled off the trees in long strands and split and dried much like Red-cedar bark. The Kwakwaka'-wakw soaked it for up to 12 days in warm salt water in a quiet bay at the low water line, then pounded it on a flat stone with a whale-bone

Yellow-cedar bark drying in Massett, Haida Gwaii.

A Yellow-cedar-bark hat made by Florence Davidson of Haida Gwaii.

Haida kerfed (bent-wood) food bowl made from Yellow-cedar, similar to kerfed Red-cedar boxes. This bowl is adorned with crest figures. (RBCM 14678)

A Yellow-cedar-bark cape. (RBCM 10806)

beater to make it soft and pliable, dried it for four days or so, and stored it. For more details on the preparation of the inner bark, see the account for Red-cedar.

People along the coast used the prepared bark for cordage and for weaving blankets, capes and other items of clothing; they preferred it to Red-cedar bark because of its softness. They often interwove or trimmed Yellow-cedar bark with duck down, Mountain Goat wool or Black Bear fur. The Chilkat Tlingit people of Alaska wove their famous Chilkat blankets with Mountain Goat wool over strands of Yellow-cedar bark. People all along the coast also used Yellow-cedar to weave mats and hats, and for decorating masks. They also shredded the bark to make bandages and "wash cloths" for babies, and to use as tinder.

Rocky Mountain Juniper (Cypress Family)   *Juniperus scopulorum* (Cupressaceae)

Other Names: Rocky Mountain Red-cedar, Red-cedar.

Botanical Description
Rocky Mountain Juniper is a densely branching tree up to 10 metres tall, but can also be a sprawling shrub less than a metre high. It has stringy reddish-brown bark and small, bluish-green scalelike leaves.

Male and female cones are produced on separate trees. The female cones are berrylike, the size of peas, and bluish-purple with a whitish waxy coating; most have one or two seeds and mature in two seasons. The foliage and fruit have a spicy, pungent odour, especially when crushed.

Habitat: rocky coastal islands, dry plains, valleys and lower mountains.

Distribution in British Columbia: dry sites from coastal Vancouver Island and the Gulf Islands eastward to the Rocky Mountains and north to the Peace and Stikine rivers.

Aboriginal Use
The wood of Rocky Mountain Juniper is extremely tough. Interior Salish peoples and the Carrier considered it one of the best materials for making bows. The Nlaka'pamux also used it to make hoops, clubs and hafts. The Okanagan made a type of spoked wheel used in a throwing game and, after horses were introduced, made double-tree yokes from juniper wood. The Secwepemc used it to make snowshoe frames and spears. The Carrier and some Interior Salish groups made rough temporary spoons from the inner bark, and the Nlaka'pamux made elongated juniper bark baskets.

Virtually every group within the range of Rocky Mountain Juniper, from the Straits Salish of Vancouver Island to the Blackfoot of Alberta and the Flathead of Montana, used the pungent boughs to clean and fumigate houses and purify the air, especially following an illness or death. They believed the scent would protect the inhabitants from bad influences. To fumigate a house, they burned the branches in the fire, or placed them on top of the stove and allowed them to smoke like incense, or simply hung them around the walls. To make a washing solution for floors, walls, bed-clothes and furniture,

Interior Salish Juniper-bark spoon.
(RBCM 2671)

they boiled the boughs in water. The Flathead used Rocky Mountain Juniper scent as a perfume. The Ktunaxa used bundles of the boughs to sprinkle water on the hot rocks during sweat-bathing.

The Nlaka'pamux used juniper wood as a fuel for smoking hides, often combining it with Big Sagebrush when they desired a dark skin. The Secwepemc noted that juniper wood should not be used for cooking because it will impart a bitter flavour to the food. The Haisla made labrets from the wood of the related Common Juniper, and the Gitxsan used juniper wood to make small bowls and for fuel. The Nlaka'pamux used a strong decoction of the berrylike cones to kill ticks on horses, and the Okanagan soaked their arrows overnight in a solution of pounded juniper branches to make a deer's blood coagulate quickly when it was hit, preventing it from running very far. The Blackfoot, and probably some British Columbia groups, used the "berries" as beads, often interspersed with Silverberry seeds. In the American Southwest, peoples such as the Navajo still make attractive necklaces and other jewellery from the seeds of another juniper species. Some California First Peoples routinely cut bow-staves from standing juniper trees, coming back years later to cut more from the same tree (as reported by P.J. Wilke, in Blackburn and Anderson 1994).

Western Red-cedar *Thuja plicata*
(Cypress Family) (Cupressaceae)

Other Names: Pacific Red-cedar, Giant Arborvitae, Giant Cedar.

Botanical Description
Western Red-cedar is a large tree, up to 70 metres tall and 4.3 metres in trunk diameter; mature trees are often fluted and strongly buttressed at the base. The bark is thin, greyish outside and reddish-brown inside, and longitudinally ridged and fissured; it is easy to pull off in long fibrous strips. The wood is light, aromatic, straight-grained and rot-resistant. The branches are flat and spraylike, and tend to spread outward; they are yellowish-green and smoother than Yellow-cedar branchlets, and after about three or four years, they turn brown and fall

off. The branchlets hang down, and the leaves are small, flat and scalelike. The numerous male cones, borne at the branchlet tips, are reddish and about 2 mm long. The female cones grow in loose clusters on the surface of the branchlets; they are ovoid, about 1 cm long, green when immature, and brown at maturity. The cones have a few scales, which spread open when dry to release laterally winged seeds.

Habitat: rich, moist to swampy soils, usually in shaded forests.

Distribution in British Columbia: a dominant tree in moist forest habitats along the coast from Vancouver Island to Alaska; also occurs on moist slopes and in valleys of the interior, north to 54°30' latitude and up to about 1,400 metres elevation.

Aboriginal Use

Of all the plants used as materials by British Columbia First Peoples, Western Red-cedar was without doubt the most widely employed and the most versatile. The light-grained wood is rot-resistant, and easy to split and work. All coastal groups used it, and to a lesser extent, so did the interior groups who lived within the range of the tree. On the coast, Red-cedar was used to the exclusion of almost all other trees to make dugout canoes, house posts and planks, totem poles and mortuary posts, and storage and cooking boxes. Coastal people also used it to make dishes, arrow shafts, harpoon shafts, spear poles, barbecuing sticks, fish spreaders and hangers, dipnet hooks, fish clubs, masks, rattles, benches, cradles, coffins, herring rakes, canoe bailers, ceremonial drums, combs, fishing floats, berry drying racks and frames, fish weirs, spirit whistles, and paddles. They especially valued the young trees for making wedges, spear handles and splints for basketry. The Interior Salish peoples used Red-cedar to make

Red-cedar-bark canoe bailer made by Pauline Joe, Coast Salish.

various types of shelters, both temporary and permanent, as well as drying racks, spear shafts, dipnet frames, river poles, salmon spreaders, paddles, drum hoops, birch-bark canoe frames, bows and arrows; and some interior peoples, such as the Okanagan, used it for dugout canoes. People everywhere considered the wood an excellent fuel, although fast burning; many thought it made a good fire for drying fish because it burns with little smoke. According to one Secwepemc woman, Red-cedar is also good for curing and smoking hides because of its low pitch content. The Kwakwaka'wakw and other groups used it to make a drill and hearth for starting friction fires. They also made torches and kindling from Red-cedar wood. Since contact with Europeans, aboriginal peoples have used it for shingles and shakes, house siding and fence posts.

Before European contact, aboriginal people rarely felled cedar trees. Instead, they harvested fallen logs or split boards from standing trees. Felling a tree was a laborious task, usually undertaken by men. They cut around the base with adzes and chisels, or sometimes burned the trunk at the bottom until the tree toppled. The Nuxalk used to bind a Rocky Mountain Maple branch around the trunk and ignite it. It would smoulder for many hours, gradually burning through the softer cedar wood until the tree fell.

Splitting off boards, house planks and even half-logs for canoes was easier, though it was still hard work. The men would hammer in a series of graduated Yew-wood or antler wedges along the grain. They carved the wood with adzes, chisels, knives and drills. Even before European contact, some iron was available to coastal peoples, but most tools were made of stone, shell, bone or horn. Nowhere else in North America was woodworking developed to such a fine art as with Red-cedar on the Pacific coast, particularly on the central and northern coast of British Columbia. Symmetry of form, neatness, and precision characterized the most utilitarian of cedar-wood objects. Paintings and relief carvings of stylized designs, usually of animal motifs, adorned many of them, and even undecorated items were polished until they shone or carved with rows of adze marks to produce aesthetically pleasing effects. The excellence and craftsmanship of aboriginal carvers of cedar is demonstrated not only by the masks and totem poles that are famous the world over, but also (and perhaps even more so) by their dugout canoes and kerfed boxes.

Canoe making was almost always the men's responsibility. It was a highly specialized occupation, and took years of training to gain the experience to make a large canoe. Choosing and felling the tree was an important part of the overall task. Once this was accomplished, the canoe maker worked on the exterior surface of the log first, aligning and shaping the outside. Then he inserted plugs into the log to a depth equalling the desired thickness of the hull. He hollowed out the inside, usually with a combination of carving and controlled burning, until the plugs became visible. At this point the roughly formed canoe was light enough to be dragged out of the woods to the beach or to a place more convenient for the carver and his assistants to work. There, the men filled the hull with water, which they heated to the boiling point with red-hot rocks. As the wood became softer and more flexible, they spread the top so that the canoe was wider across than the original tree trunk. They sewed thwarts into place, and on more elaborate canoes, added separate prow and stern-pieces to make the craft more seaworthy. Finally, they polished the hull with dogfish skin or horsetail, and decorated it. Canoes ranged in size from those built for one or two people to those used on trading or war expeditions, some of which could accommodate 60 people. These large craft were built mainly during the 18th and 19th centuries, when iron tools were plentiful. Dugout canoes are still being made in some communities, and there have been several well-celebrated voyages up and down the coast in recent years.

Kerfed cedar boxes – also called bent-wood boxes – were a superb combination of artistry and utility. Groups of the central and northern coast, including Nuu-chah-nulth, Kwakwa̱ka'wakw, Nuxalk, Haida, Tsimshian and Tlingit, made them in many different sizes and shapes, all by the same basic technique. The box maker started with a wide, thin board split from a cedar log. He cut three transverse kerfs (grooves), spaced according to the box's ultimate dimensions. To soften the wood for bending, he steamed it over a fire covered with wet moss or seaweed, or soaked it in a creek for several days, weighing it down with rocks. When the board was pliable enough, he bent it carefully along the kerfs to make a square or rectangular box, and pegged or sewed the ends together, usually with spruce or cedar root. The box maker then attached a flanged board to the bottom with pegs and sewed it tightly enough to make the box watertight. Finally, he fitted the box with a lid. Bent-wood boxes

Haida kerfed (bent-wood) box made from Red-cedar. (RBCM 6616)

were used for boiling or steaming food, or for storing berries, fish, oil and other products. More elaborate boxes, for higher-class people or for storing ceremonial objects, were beautifully carved and decorated, although Kwakwaka'wakw elder Chief Adam Dick recalled that the boxes his family had were undecorated and dark coloured. Some types of folded boxes were used for cradles, others for coffins and still others for feast dishes and bowls (see the photograph of a Yellow-cedar bowl on page 68).

Those used for storage in ocean-going canoes were made with angled kerfs so that they were wider at the top than at the base. In the past, cedar boxes were widely traded along the coast and even into the interior. Many people still own them today. The art of making them, once virtually forgotten, is being revived by a few modern aboriginal craftsmen. In Massett, on Haida Gwaii, a small company makes beautifully crafted and painted boxes, and sells them widely.

Many coastal First Peoples made rope from the slender, pliable branches (withes) of Red-cedar. They usually gathered cedar withes in the spring and peeled off the bark. They split the thicker withes in halves or quarters and left the thin ones whole. The withes had to be twisted and worked until soft and pliable, then they could be used alone or plaited together. Sometimes, they could be softened and plaited with other withes while still attached to the tree, then cut off when they formed a rope. People used cedar-withe ropes for sewing wood, binding implements, tying boards onto house frames, constructing fences, or for anchor lines, harpoon lines, tree-climbing belts, fishing lines, fish-nets and duck nets. Some of the anchor ropes of plaited or twisted withes used by the Vancouver Island Salish were reportedly as thick as a man's wrist. A number of groups, including Nuu-chah-nulth, Kwakwaka'wakw, Comox, Nuxalk and Lower Stl'atl'imx, made baskets of cedar withes, particularly open-weave baskets for clams and other shellfish, and the Nisga'a and others employed cedar withes to make fish traps. The Squamish used them for tying bundles of dried salmon. Some Coast Salish groups, such as the Nanaimo and Saanich of Vancouver Island, used bundles of dried cedar twigs

Coastal-style cedar-withe basket.
(RBCM 6589)

as torches. Thin strips of cedar wood, cut from sapling trees and split along the annual rings, were sometimes used in basketry with cedar withes or roots.

Some peoples, such as the Secwepemc and Squamish, used cedar boughs for bedding. The Saanich, Comox and other coastal groups anchored bunches of branches in the ocean near the shore to collect herring spawn, although Western Hemlock boughs were more commonly used for this purpose. The Comox threaded Eulachons and herring onto green withes for drying and the Kwakwaka'wakw interspersed the boughs with layers of edible seaweed being dried. The Nlaka'pamux reportedly used them to make a green dye.

Cedar roots were used by coastal groups, such as the Nuu-chah-nulth and Kwakwaka'wakw, for lashings and for making nets, baskets, hats and mats, but the Salish people, especially those of the interior, were by far the greatest users of the roots. The coiled split-cedar-root baskets of these peoples, particularly the Lower Stl'atl'imx and Lower Nlaka'pamux, are world famous. The foundation coils of these baskets were made of inner cedar bark, cedar-root bark, bundles of split cedar root or thin splints of cedar sapwood, and were completely covered and, at the same time, stitched tightly together by strands of split cedar root. So closely were they sewn that the baskets were watertight, serving equally well as berry containers, water carriers or cooking vessels. Basket makers decorated their baskets by a process known as imbrication, in which strips of material such as Bitter Cherry bark (naturally red or dyed black) and Reed Canary Grass stalks are superimposed over the basic cedar root to produce beautiful geometric designs and patterns of plant and animal motifs. Cedar-root baskets varied considerably in size and shape, the most common style being rectangular with a small, flat bottom, steeply flaring sides and rounded corners. Even baby carriers were often made of split cedar root. People usually dug up the roots for these baskets in spring, heated them in a fire, peeled, split and bundled them for later use. Bundles of cedar root were a common trading item among Interior Salish peoples. The Secwepemc of the Canim Lake area used to dig the roots in the mountains northeast of

Coiled cedar-root basket materials.

Left: Nellie Peters of Mount Curry (Stl'at'imx) holding a baby's cradle and baskets made of split cedar root. The basketry is decorated with dyed black and undyed (reddish) cherry bark.

Below: Stl'atl'imx cedar-root basket. (RBCM 14614)

the lake, process them and then sell them at the annual Secwepemc tribal gathering at Green Lake. Peoples such as the Secwepemc also constructed cedar-root fishnets, and used the roots to stitch together birch-bark baskets and canoes.

The whole bark of the Red-cedar tree is another very important material. People removed it in long sheets or slabs, fattened it and used it, without further modification, as roofing and siding for temporary shelters. The Interior Salish used these sheets as a covering for sweat-houses, as flooring for houses and canoes, and as a lining for underground food caches. The Secwepemc made rough cooking vessels for boiling fish by cutting off a piece of bark about a metre long, folding it and tying it around the top with a twisted Red-osier Dogwood branch. The Cowichan and other groups made similar vessels. The Okanagan sometimes used the bark to insulate their Tule tipis.

The Sechelt placed cedar bark between layers of edible roots, including potatoes and turnips, to prevent them from rotting in storage. Many coastal peoples made canoe bailers from sheets of cedar bark. The Haida threaded Salmonberry sticks at intervals across the inner surface of large sheets of the bark to keep them flat. Then they piled them up and weighted them down with stones. They also used strips of bark for roofing and sold them to their mainland neighbours along the

Nass River – the price at one time was one blanket for two sheets of bark. The Cowichan used cedar bark steeped in water for tanning fish hooks.

The most valuable part of Red-cedar bark was the fibrous inner portion. It was used by virtually every group that had access to the tree, but especially by coastal peoples. They split the inner bark into strips for weaving open- and closed-work baskets, bags, hats, mats (for walls, floors and mattresses), capes, and blankets (although Yellow-cedar bark was usually preferred for the last two items). They carefully pounded and twisted it into string to make shaman's and dancer's ceremonial head rings, neck rings, armbands and belts, to make fishing lines, ropes, harpoon lines, animal snares, and nets, and for threading clams and fish for drying. They used finely shredded inner bark to decorate masks, to make brooms, paint brushes, work aprons, skirts, capes and dance costumes, to use as tinder, napkins, towelling, bandaging, diapers and infant bedding, and to cover the hands of drummers during winter dances. Some coastal peoples made a "slow match" by binding a bundle of shredded cedar bark tightly with some less combustible substance. Once ignited, the inner core would smoulder inside its wrapping for many hours. The Saanich made torches from the inner bark of Red-cedar. The Fraser River Sto:lo spun it together with Mountain Goat wool, dog hair and bird feathers. In some areas people used larger pieces of the inner bark to make canoe bailers, spoons and storage bags.

Each group had its own techniques for harvesting and preparing Red-cedar bark, but the general method remained the same. The harvester, usually a woman, selected a fairly young tree, about a third of a metre in trunk diameter, with few lower branches. Women often travelled long distances along creeks and into the mountains to find trees with suitable bark for harvesting. When she found a good tree, the woman made a transverse cut through the bark, a metre or so from the ground, about one-quarter or one-third the circumference of the tree. She pried up the bark at the bottom and along the edges, being careful not to split it too much, then holding the bottom tightly, she pulled outward and upward. As it was pulled off the tree, the bark strip became narrower towards the top until it broke off. A good bark harvester could often pull off a strip 9 metres or more in length. Once she had the bark off the tree, she peeled away the brittle outer bark and either left it behind or took it home for fuel. The inner bark is about 5 mm thick, with a leatherlike texture and light colour. Each strip could be folded into an elongated bundle and tied neatly with the thin upper end, ready for

Florence Davidson of Massett harvesting Red-cedar
bark. Holding tight (left), and pulling up and out . . .
until (right) the long strip comes off near the top.

transport home. It might take one or two long strips to make a single
mat or basket.

At home, the woman split the fresh bark into two thinner sheets –
the innermost sheet was considered higher quality. At this stage, the
bark could be dried for later processing, or could be split and woven
while still fresh. Most people shredded and softened the fibres by
pounding the bark over the edge of a paddle or board with a beater of
Yew wood, stone or whale bone. Sometimes they softened the bark by
immersing it in fresh or salt water for several days. Dry bark fibre could
be rendered flexible again simply by soaking it. Before weaving a basket, hat or mat, the weaver split off long strips of bark and bundled
them separately so they would be ready to use as needed. The natural
colour of the bark when fully aged is dark reddish brown. It could be
dyed red with the bark of Red Alder, or black with charcoal or by im-
mersion in water in which rusty nails had been soaked. Basket makers
interwove coloured and undyed strands of bark to produce patterns and
special effects. People also used dyed shredded bark for ceremonial
head and neck rings and for decorating masks. In a recent interview,
Kwakwaka'wakw historian Daisy Sewid-Smith explained the significance of cedar-bark ceremonial regalia, as it relates to the origins of her
people (Sewid-Smith and Dick, in press).

The Western Red-cedar and its relative, Yellow-cedar, are featured
in the comprehensive and well-illustrated book, *Cedar: Tree of Life*, by
Hilary Stewart (1984).

Note

Harvesting cedar bark is harmful to the tree. Even when the trunk is not completely girdled, the wood is exposed and the capacity of the tree to transport nutrients is reduced. Aboriginal bark collectors are careful not to overharvest bark from any tree, and evidence of their sustainable harvesting methods can be seen in many old-growth forests – these are known as Culturally Modified Trees. Do not try to harvest cedar bark unless you know the harvesting protocol.

Tsimshian initiate's collar made of Red-cedar bark, natural colour and dyed with Red Alder bark. (RBCM 1537B)

Grand Fir	*Abies grandis*
Amabilis Fir	*A. amabilis*
(Pine Family)	**(Pinaceae)**

Other Names: White Fir, Balsam Fir (*A. grandis*); Pacific Silver, Silver Fir, Cascade Fir, Lovely Fir, Red Fir (*A. amabilis*).

Botanical Description

Both Grand Fir and Amabilis Fir are true firs. They are tall, straight trees up to 70 metres tall. When young, the bark is smooth, light grey to silvery, and thin, with prominent pitch blisters. As the bark ages, it becomes thicker, rougher and often lighter coloured. The needles are dark green above, white on the lower surface, and notched at the tips. Most Grand Fir needles are 2 to 4 cm long and – especially on young trees – are twisted to lie in a horizontal plane, giving the branches a characteristic flat, spraylike appearance. The needles of Amabilis Fir

Grand Fir.

are similar, but this tree has an additional set of shorter needles that grow along the tops of the twigs, and lie close to the twigs, pointing forward. The pollen cones of both species are reddish and hang from the undersides of the upper branches. The seed cones, borne near the top of the tree, are cylindrical, 6 to 11 cm long, stiffly erect and often quite pitchy; they shed the scales and bracts with the seeds in the fall, leaving only the central axis on the branch. The seed cones of Grand Fir are light green, and those of Amabilis Fir are purplish. The seeds of both species are light brown and have large, membranous wings.

Habitat: Grand Fir grows in damp to moderately dry coniferous forests; Amabilis Fir is found in moist, rich soils in shady bottomlands and montane forests.

Distribution in British Columbia: Grand Fir is found from sea level to moderate elevations on Vancouver Island and the adjacent mainland, recurring in the valleys of Kootenay Lake and the Arrow Lakes in the interior wet belt; Amabilis Fir grows from sea level to subalpine elevations along the Pacific coast from Alaska to Vancouver Island, extending inland along some of the major river valleys. Neither species occurs on Haida Gwaii.

Aboriginal Use
The wood of these firs is soft and rather brittle, so aboriginal peoples did not use it often. Still, some groups found uses for it. The Straits Salish, Ditidaht and Kwakwaka'wakw made hooks for halibut and dogfish from Grand Fir knots. They obtained the knots – dense, pointed branch-ends – from rotten logs. They split the knots lengthwise, then steamed and bent the pieces into a horseshoe shape. The Nisga'a sometimes split house planks from Amabilis Fir trees. The Gitxsan used the wood of Amabilis Fir as fuel, whereas the Chehalis of Washington, and probably some British Columbia groups, used Grand Fir. The Okanagan of the Arrow Lakes region made canoes from sheets of Grand Fir bark, and rubbed the pitch on the backs of bows after they had been wrapped with Bitter Cherry bark to give them a good grip.

The Straits Salish of Vancouver Island made a brown dye for basketry from Grand Fir bark and a pink dye by combining it with red ochre; they also rubbed the pitch on canoe paddles and other wooden articles, then scorched them to give a good protective finish.

The boughs of both species have a pleasant, spicy citruslike fragrance. The Nuu-chah-nulth and Nlaka'pamux, and probably other groups, used the boughs as bedding and floor coverings, and for covering berry baskets. The Kwakwaka'wakw used Grand Fir and probably Amabilis Fir branches for ritual scrubbing in purification rites, and shamans made costumes and head-dresses from them; sometimes they also rubbed the pollen over their bodies. The Haisla used the boughs to line Eulachon-ripening pits and also as a "sled" to pull cargo over deep snow. The Flathead of Montana dried and pulverized the needles and used them as a baby powder and body scent.

Subalpine Fir
(Pine Family)

Abies lasiocarpa
(Pinaceae)

Other Names: Alpine Fir, Balsam, Balsam Fir, White Balsam Fir, "Sweet Pine".

Botanical Description
Subalpine Fir is a small tree, usually less than 30 metres tall, with a narrow, spire-shaped crown and short, thick branches. The bark is smooth and ashy-grey. The needles are bluish-green, flat and 2 to 3.5 cm long; needles on the lower branches are blunt, and those on the upper branches are pointed – all are crowded and tend to turn upward. Stomata – the small gas-exchange pores on the needles – grow on both surfaces of the needles, in a single whitish median band on the upper side and two lateral bands on the

underside. The pollen cones are small and bluish. The seed cones, borne near the top of the tree, are erect, 6 to 10 cm long, and usually deep purple, becoming lighter with age; they shed their scales with the seeds in the fall, leaving only the central axis. The seeds are light brown with prominent wings. Some botanists classify *Abies lasiocarpa* as *A. balsamea* subsp. *lasiocarpa.*

Habitat: moist subalpine forests and open slopes at or near the timberline, where the growth is usually considerably stunted.

Distribution in British Columbia: montane forests from the coast to the Rocky Mountains and from Washington north to Alaska, descending to 600 metres in the central and northern interior, common from 1,100 metres to the timberline in the south; not found on Haida Gwaii.

Aboriginal Use
The Carrier made roofing shingles from the soft, even-grained wood of Subalpine Fir. They also burned the rotten wood as a smudge for tanning hides and probably as a general fuel. And they rubbed Subalpine Fir pitch on the seams of birch-bark canoes and coated their bowstrings with it. The Haisla of the Kitlope Valley used the wood to make chairs and special insect-proof storage boxes for dancing regalia. The Gitxsan made snowshoes from young Subalpine Fir wood. Some people chewed the pitch of Subalpine Fir and other coniferous trees to clean their teeth. The Secwepemc, and probably other interior groups, made large temporary baskets from sheets of Subalpine Fir bark. These baskets were barrel- or funnel-shaped, with the exterior surface of the bark forming the inside of the basket and the seams stitched with spruce or cedar root. Secwepemc people used these bark vessels for cooking berries or soaking skins, but they did not consider them as high quality as birch-bark baskets. The Gitxsan used sheets of the bark for roofing. Many peoples in the province used the boughs as bedding. The Secwepemc pushed the broken ends of the branches into the ground to produce a soft, springy sleeping surface. People also used the boughs for flooring in the sweat-house, for standing on after swimming and as a surface for butchering deer.

The Flathead of Montana placed the fragrant needles on the stove or hung them around the walls of a house to clean and purify the air. They dried and pulverized the needles to make a baby powder and a body and clothing scent. They also mixed them with lard and applied them to the hair as a perfume and green tint. The Blackfoot of Alberta burned the needles as incense at ceremonials, and used them as a sachet.

Nlaka'pamux elder Annie York stated that people venturing into the mountains – especially young women – would rub their skin with Subalpine Fir boughs to mask their human scent and reduce the risk of being attacked by large predators, such as bears. Secwepemc elder Mary Thomas noted that Subalpine Fir pitch makes an effective insect repellent.

Western Larch   *Larix occidentalis*
(Pine Family)   (Pinaceae)

Other Names: Tamarack, Western Tamarack, "Red Fir".

Botanical Description
Western Larch is a large, handsome tree up to 70 metres tall, with thick, flaky cinnamon-coloured bark; a mature tree has few branches on its lower trunk. The pale green needle-like leaves are 4 to 5 cm long and grow in clusters of 15 to 30. Unlike most conifer needles, the leaves of Western Larch are deciduous, turning golden yellow and dropping in autumn. The yellowish pollen cones are about 1 cm long, and the seed cones 3 to 4 cm long, at first purplish-red, later reddish-brown.

Western Larch with Black Tree Lichen hanging from its branches.

Habitat: mountain valleys and lower slopes, often in swampy areas, usually in mixed stands.

Distribution in British Columbia: on north slopes and in higher valleys of the Kootenay, Arrow, and Okanagan drainage systems, from the United States border north as far as Shuswap Lake.

Aboriginal Use

The wood of Western Larch is rather difficult to work – apparently, the First Peoples of British Columbia seldom used it. The Nlaka'pamux and Okanagan heated the pitch in a fire until it was hard and dry, then pounded it with a stone or a mortar and pestle into a fine reddish powder. By mixing this powder with Cottonwood bud resin to form a sticky paste, they made a general purpose red paint for colouring wood, buckskin and other materials. Girls aged 9 to 16 would mix the powder with grease and smear it on their faces as a cosmetic and skin conditioner. From descriptions of the locations of some of the trees used – between Hope and Princeton in the vicinity of Manning Park – it is possible that these groups also used the pitch of a related species, Alpine Larch. The Dena'ina of Alaska used Subalpine Larch wood, which is relatively hard, for boat ribs and sled runners.

According to Okanagan tradition, when larch trees turn yellow in the fall, it is a sign that pregnant female bears are retiring to their dens for the winter; it was also believed that if larch needles fell on the rump of a pregnant bear she would have a miscarriage.

Engelmann Spruce	*Picea engelmannii*
White Spruce	*P. glauca*
(Pine Family)	(Pinaceae)

Botanical Description

Engelmann Spruce is a straight, spire-shaped tree up to 50 metres tall. White Spruce is usually less than 25 metres tall, and is often deformed and stunted. The bark of Engelmann Spruce is brownish-red and scaly; that of White Spruce is silvery-brown and also scaly. The needles of both are bluish-green, slender, sharp and four-sided, tending to project from all sides of the twigs, or commonly turning upward. The pollen cones of Engelmann Spruce are yellow, those of White Spruce reddish. The seed cones of Engelmann Spruce are 4 to 6.5 cm long, yellowish-brown to purplish-brown, the scales thin with wavy margins; those of White Spruce are 3 to 5 cm long, light brown to purplish, the scales usually stiffer, blunter and more regular than those of Engelmann Spruce. These two species are closely related and commonly hybridize in areas where their ranges overlap. They are often treated as geographic races of the same species, *Picea glauca*: Engelmann Spruce as ssp. *engelmannii* and White Spruce as ssp. *glauca*. To distinguish between them is very difficult – the shape of the cone scales and the different ranges are probably the best indicator.

Engelmann Spruce.

Habitat: Engelmann Spruce grows in damp, shady subalpine forests, in upland swamps and on alluvial floodplains; White Spruce generally occurs in damp boreal forests, swamps and floodplains, usually at lower elevations but higher latitudes than Engelmann Spruce. Both species are highly frost resistant, White Spruce even more so than Engelmann.

Distribution in British Columbia: Engelmann Spruce occurs mainly east of the Coast and Cascade mountains to the Rockies, south of latitude 51°31'N and above 900 metres elevation, but does appear sporadically further north at lower elevations; White Spruce is widespread in the northern interior forests, extending south to the Interior Plateau and eastward into the Rocky Mountain Trench.

Aboriginal Use

Aboriginal people have seldom clearly distinguished between these spruce species. It seems likely that the groups of the southern interior commonly used Engelmann Spruce, while northern interior peoples used White Spruce.

The Tahltan used the wood from spruce saplings to make snowshoe frames, but considered it somewhat heavy and brittle, and not durable. They also used it for gambling sticks and sometimes for bows. Tahltan and other peoples heated spruce gum and used it to glue skin onto bows and arrowheads onto shafts. They also burned the decayed wood in the process of tanning hides. The Carrier used a type of spruce wood to make the encircling pieces for fish traps, but this may have been Black Spruce, which they sometimes employed for snowshoe frames and drying poles. The Dena'ina of Alaska used Sitka, White and Black spruces for many purposes, as described under Sitka Spruce.

People throughout the interior used spruce roots, peeled, split and soaked in water, for sewing the seams and rims of birch bark and other

Tsilhqot'in coiled spruce-root basket. The spruce root is only visible on the inside. The entire outside of the basket is imbricated with grass (the light-brown background, possibly Giant Wildrye) and cherry bark (the darker designs). The rim is decorated with bird quills. (RBCM collection)

types of bark baskets. The Interior Salish and Athapaskan peoples used them, like Red-cedar roots, to make tightly woven coiled baskets; in fact, the Carrier, Tsilhqot'in and some Secwepemc used spruce roots even more than cedar roots for this purpose.

Many interior peoples used sheets of spruce bark to make large cooking baskets, similar in style to birch-bark baskets, but not as well finished. They stitched the seams and tops with spruce root. The Nlaka'pamux, Secwepemc and Ktunaxa also used spruce bark to make canoes, the Nlaka'pamux used it for roofing, the Stl'atl'imx for baby carriers and as covering for summer lodges, and the Carrier to make trays for gathering and processing berries.

Secwepemc hunters made bedding from spruce boughs, covering them with a thick layer of fir branches as protection against the sharp foliage. Secwepemc children learned that if they became lost to always look for a spruce tree for shelter; the thick, hanging branches afforded considerable protection from the elements. Carrier people, when camping, used to strip the spruce needles from the boughs to make a floor covering for tents.

Carrier spruce-bark basket. (RBCM 9262)

Sitka Spruce
(Pine Family)

Picea sitchensis
(Pinaceae)

Botanical Description
Sitka Spruce is a large tree, up to
70 metres tall and 2 metres or
more in trunk diameter. The bark
is thin and greyish with rounded
deciduous scales about 5 cm
across. The branches are characteristically droopy. The needles
are stiff, sharp-pointed, diamond-
shaped in cross-section, about 2.5
cm long and tend to project from

all sides of the twig. The cones hang down from the branches; they are
pale brown, about 6 cm long and cylindrical, with pale brown, papery
scales. In the northern part of its range Sitka Spruce often hybridizes
with White Spruce.

Habitat: humid west-coast forests, from sea level to about 600 metres
elevation.

Distribution in British Columbia: confined to a coastal strip about
80 km wide; especially common on Haida Gwaii and on the west coast
of Vancouver Island.

Aboriginal Use
Sitka Spruce wood is light and strong. First Peoples often used it to
make pegs for cedar boxes. The Kwakwaka'wakw and their neighbours
sometimes used it to make digging sticks, herring rakes, arrows, bark
peelers and slat armour. The Haida used both wood and bark for fuel.
These groups, and probably others along the coast as well, made cod
and halibut hooks from spruce knots and branches, first steaming them
to make them easier to mould into hooks. But Ditidaht elder John
Thomas explained that spruce knots did not hold their shape as well as
those of Western Hemlock, fir or Cottonwood. The Gitxsan used
spruce for firewood, gaff handles and the wings of salmon traps.
   The Makah of Washington and the Dena'ina of Alaska made wedges
from the wood of larger Sitka Spruce roots. The Nisga'a and Haida
sometimes split house boards from Sitka Spruce trunks. To produce

extra wide boards for the houses of wealthy and important people, they cut curved pieces from the circumference of the largest trees, then steamed them and weighted them down until flat. The Dena'ina also used spruce (Sitka, White and Black) for sleds, spear shafts, digging sticks, shovels, fire-making drills, tongs, drums, dishes, mauls, bows, dugout canoes, fish traps, clubs and may other items. They distinguished between the dark hard wood of slow-growing trees and the light wood of fast-growing trees. They also used the bark to dye their fishnets, and used sheets of bark for roofing, flooring and siding for shelters.

The Nuu-chah-nulth, Kwakwaka'wakw and other peoples used spruce gum to cement the joints of implements such as harpoons and spears and for caulking canoes. Before hunting for porpoises, the Kwakwaka'wakw greased their canoes with spruce-scented tallow to mask human odours. Coastal people also used spruce gum on the ends of torches.

Sitka Spruce roots were an important basket material among coastal peoples. They were also used for making nets and for sewing wood (such as in box construction), and as a single or multi-strand rope for tying on house planks, binding fish hooks and harpoons, and for fishing and harpoon lines. The Nisga'a used them to stitch the rims of their birch-bark baskets. For basketry, women harvested spruce roots in early summer. They picked trees with few lower branches growing in sandy soil; trees growing in rocky soil had twisted and knotted roots. Root harvesters took great care in selecting the longest, straightest roots, and chose different sizes depending on the intended use. They bundled each root separately and scorched it briefly over a hot fire,

Florence Davidson and her grand-caughters show the author (left) how to harvest spruce roots, near Massett on Haida Gwaii.

Florence Davidson (left) and her sister processing spruce roots.

enough to heat the wood of the root, but not to burn through the bark. Any roots not heated correctly would soon turn brown. To remove the bark, the woman pulled a root through a split stick or an upright set of fire tongs. Then she carefully split it along the vertical plane where the line of small secondary rootlets appeared. Depending on the size of

Haida spruce-root hat. (RBCM 154)

the root and its potential use, she might split each half again, yielding two flat strands and two rounded ones. These she pulled over a knife blade or some other flat object until they were completely flexible, then she dried them. The root strands could be made workable again simply by soaking them.

Most coastal groups made spruce-root baskets, both closely twined and of more open weave. But the northern people – especially the Haida and the Tlingit of Alaska – were exceptionally skilled in making them. So finely and tightly did they twine the roots that the baskets could be used to pack water, and when held to the light not a leak could be seen. Watertight spruce-root hats were another specialty of these peoples – and they still are. The Haida decorated their baskets with false embroidery or by substituting dyed wefts for naturally coloured ones. The Tlingit decorated their baskets with false embroidery. Both peoples painted designs on their hats using charcoal or natural mineral paints, an activity usually undertaken by men.

The Kwakwaka'wakw used spruce branches to make rope. Some people used the boughs to make shelters when camping – the prickly needles discouraged animal intruders. The Gitxsan sometimes used spruce bark (mostly hybrid spruce) for roofing if they could not obtain cedar bark for this purpose.

Lodgepole Pine
(Pine Family)

Pinus contorta
(Pinaceae)

Other Names: "Jack Pine", Black Pine, Scrub Pine, Shore Pine, "Red Pine".

Botanical Description

Lodgepole Pine is a small to medium tree, rounded or pyramidal, 10 to 30 metres tall, with thin limbs often crowded at the top third of the tree when it is growing in dense stands. The bark is thin, rough, reddish-brown to greyish-black, and scaly.

The needles, in pairs, are 3 to 6 cm long, often with a yellowish-green tinge. The pollen cones are small and reddish-green, and grow in clusters. The woody seed cones are oval-shaped and usually slightly lopsided, tending to remain closed on the tree for many years, some opening only after a hot forest fire.

Habitat: a highly adaptable species, found from coastal dunes and bogs to rocky hilltops and upland plains, often forming pure stands, especially after a fire.

Distribution in British Columbia: widespread throughout the province from sea level to subalpine elevations; especially common as a successional species in burned-over forests east of the Coast and Cascade ranges.

Aboriginal Use

Lodgepole Pine wood is soft, light and straight-grained, but it is weak and not very durable. As the name suggests, this tree was commonly used for housing. The Okanagan and Nlaka'pamux used Lodgepole Pine trunks to make tipi poles and other types of building materials, as did the Flathead of Montana and the Blackfoot of Alberta. They usually cut the poles in summer and immediately debarked, trimmed and sun-dried them. The Stl'atl'imx have recently used the trunks to build log cabins. The Kwakwa̱ka̱'wakw and their neighbours used the wood to make fire tongs, board protectors for bending boards, cedar-bark peelers, digging sticks, maul heads and harpoon shafts; the Tahltan sometimes made arrow shafts from the wood. The Gitxsan used the poles for manoeuvring their canoes, and split poles in their salmon traps. Lodgepole Pine wood was a major source of fuel for hunters and travellers; because it is very pitchy, the Secwepemc said it would burn even when green. The Carrier made fire-making drills from it, and the Stl'atl'imx and Gitxsan used it to make torches. The Tahltan burned the decayed wood as a smudge for tanning hides. The Haisla used burning pine twigs to trim their hair and pine charcoal as a pigment for tattooing.

The long, straight trunks of Lodgepole Pine give this tree its name.

Aboriginal peoples throughout the province also used other parts of the tree. The Nisga'a occasionally split and twisted the roots for rope. The Haida employed sheets of the bark as splints for broken limbs, and the Stl'atl'imx sometimes used them for covering summer lodges. The Tahltan strewed the boughs on the floors of their houses.

Lodgepole Pine pitch is a highly versatile substance. It was used by the Sechelt to waterproof canoes and baskets and to make a hot fire, and by the Saanich and others to glue arrowheads onto shafts. The Lower Stl'atl'imx sealed fish hooks with pine pitch; they also mixed it with bear grease and heated it to make a protective coating for Indian Hemp fishing line and to glue Bitter Cherry bark over the joints of harpoons and other implements. The Nlaka'pamux also mixed it with grease and rubbed it on the outside of stone pipes to give them a glossy finish.

White Pine
(Pine Family)

Pinus monticola
(Pinaceae)

Other Name: Western White Pine.

Botanical Description

White Pine is a slender, attractive, medium-sized tree, growing 30 to 60 metres tall. The bark, thin and greyish when young, becomes thicker and breaks into rectangular flaking scales with age. The needles grow in bundles of five; they are bluish-green, slender and 5 to 10 cm long.

The pollen cones are yellowish and clustered, usually under 1 cm long. The seed cones hang from the upper branches; they are cylindrical, 15 to 25 cm long and 6 to 9 cm thick at maturity. The scales are green to purplish when young, turning brownish with age, and often tipped with large globules of white pitch. The seeds are 7 to 10 mm long, with wings two to three times as long.

Habitat: well-drained, sandy soils in valley bottoms to open slopes up to elevations of 1,800 metres on the coast and 1,100 metres in the interior.

Distribution in British Columbia: on Vancouver Island and the adjacent mainland, east to Manning Park in the Cascades; common in the interior wet belt, south to the American border and north to Quesnel Lake.

Aboriginal Use

White Pine wood is light, moderately strong and durable, but was seldom used in carving or construction. But the Manhousaht Nuu-chah-nulth used it to make long needles for mat making, and reportedly named the tree after this use.

The Secwepemc, Ktunaxa and Arrow Lakes Okanagan peeled off the bark in large sheets and used it to make storage baskets and small canoes; so did the Stl'atl'imx and the Skagit of Washington, though rarely. George Dawson (1891) reported that, as of 1891, pine-bark canoes were still occasionally used on Shuswap Lake near the Columbia River. As with birch bark, the inner side of the bark became the outside of the canoe. The seams were sewn with roots and the inside was strengthened with wooden ribs and thwarts which were lashed in place. Knot-holes and fissures were plugged with resin. Pine-bark canoes were said to be very swift and, when balanced properly, remarkably seaworthy. People made pine-bark baskets in the same fashion as those of birch bark, stitching them with roots and strengthening the tops with a twisted withe rim. The Sechelt sometimes used White Pine pitch for waterproofing.

Ponderosa Pine
(Pine Family)

Pinus ponderosa
(Pinaceae)

Other Names: Yellow Pine, Red Pine, Bull Pine.

Botanical Description

Ponderosa Pine is a large forest tree, commonly 30 metres or more tall, with thick, reddish, furrowed bark that flakes off in irregular scales. The yellowish-green needles, usually growing in threes, are longer than those of any other conifer in British Columbia, frequently exceeding 20 cm. They are usually clustered to-

wards the branch ends, giving the tree a feathery appearance. The pollen cones are yellow to purple and strongly clustered; the female or seed cones are broadly oval-shaped, reddish-purple when young, and brown when they mature after two years. The seeds are 6 to 7 mm long, with prominent wings to aid in dispersal.

Habitat: dry, warm valleys and slopes up to 900 metres; intolerant of shade and extreme cold.

Distribution in British Columbia: forming open parklike forests in the dry southern interior of British Columbia east of the Cascade Mountains, from the Fraser and Thompson river canyons to the Okanagan and Similkameen valleys, as far north as Clinton in the Cariboo; recurring in the dry sections of the Kootenay and Columbia river valleys.

Aboriginal Use

The Southern Okanagan and Fraser River Nlaka'pamux often made dugout canoes from Ponderosa Pine logs. The wood is heavier than cedar wood, and is said to split more easily. The Okanagan also used it for cache poles. All southern interior peoples valued the wood, bark and cones as fuel. The Okanagan used the wood as a hearth for making friction fires, and made the drill from a dried-out section of the leading shoot from the last year's growth. The Secwepemc burned the wood, bark and pitchy tops of Ponderosa Pines on camping trips, because they burned fast and cooled quickly, making it difficult for enemies to tell how long ago the camp had been broken. They and the Okanagan used the rotten wood for smoking hides, while the Nlaka'pamux used the cones for this purpose, often mixing them with Douglas-fir bark. The Okanagan used the spiny, immature cones for "combing" out the fibres of Indian Hemp stems. They also rubbed finished Indian Hemp fishing lines with Ponderosa Pine gum to help preserve them, and did the same to other fishing implements bound with Indian Hemp twine.

Ponderosa Pine boughs have a spicy fragrance and were used generally as bedding and for covering floors. They were especially useful in doorways and on paths during the winter because they promoted rapid melting of the snow. The needles were used as insulation for cellars, food caches and underground storage pits, and when dry made good tinder. The Secwepemc used to gather the pollen in springtime, mix it with hot water and use this concoction to colour clothing a light yellow.

Douglas-fir
(Pine Family)

Pseudotsuga menziesii
(Pinaceae)

Other Names: Oregon Pine, simply "fir" to many people.

Botanical Description

A giant forest tree, Douglas-fir grows up to 70 metres tall on the coast, but seldom more than 40 metres in the interior. The bark of young trees is smooth and grey-brown, often with resin blisters; but on old trees it becomes thick and furrowed, grey outside and mottled red-brown and whitish inside. The needles are flat,  pointed but not prickly, 2 to 3 cm long, uniformly spaced along the twig and spreading from the sides and top. The pollen cones are small and reddish-brown; the seed cones, which hang from the branches, are green when young, and when mature turn reddish-brown to grey and fall off. Prominent three-pointed bracts extend well beyond the cone scales. Cones of coastal trees are usually 6 to 10 cm long, while those of interior trees are 4 to 7 cm. The seeds are 5 to 6 mm long, with prominent wings. Douglas-fir has two well-defined geographic races in British Columbia: coastal Douglas-fir is *Pseudotsuga menziesii* var. *menziesii* and Interior Douglas-fir is *P. menziesii* var. *glauca*.

Habitat: moist to very dry areas, from sea level to 1,500 metres in the southern interior, and as high as 1,800 metres in the Rockies.

Distribution in British Columbia: widespread throughout the southern half of the province, extending as far north as Stuart and McLeod lakes and up the Parsnip River in the interior and north as far as the Kitlope River valley on the coast. Does not occur on Haida Gwaii or most of the central and northern coast.

Aboriginal Use

Douglas-fir wood is heavy, strong, fine-grained and durable. It seasons well and is easy to work. The Okanagan used it to make tipi poles, drying scaffolds, smoking racks and spear shafts; and the Secwepemc for

A Nlaka'pamux man fishing on the Fraser River with a Douglas-fir dipnet.

spear shafts, gaff-hook poles, canoe thwarts and river poles, although they preferred cedar for these items. Okanagan ice fishermen blackened their Douglas-fir spears in the fire to make them invisible to fish. One Secwepemc man reportedly made a fir-wood dugout canoe that lasted for many years. The Carrier, Secwepemc and Ktunaxa used Douglas-fir to make snowshoes; the Carrier also made fish traps with it. The Stl'atl'imx used forked Douglas-fir saplings to make dipnet frames, moulding them to a circular shape while still green. They also made harpoon shafts from the branches, as did the Katzie (Sto:lo), Squamish and various other Coast Salish groups. Coast Salish and other peoples frequently used fir wood to make spoons, spear shafts, dipnet poles, herring and Eulachon rakes, harpoon barbs, fire tongs, and salmon weirs. On Vancouver Island, and perhaps the mainland, the Salish people moulded halibut and cod hooks from fir knots, as well as those of Western Hemlock, by steaming them, placing them in a section of Bull Kelp stipe overnight to give them the right curvature, then drying them and rubbing them with tallow to waterproof them. The Kwakwaka'wakw sometimes made coffins of fir wood. The Comox prepared dogfish flesh by stuffing it with powdered rotten Douglas-fir wood and burying it for a period of time in a pit lined with the same material.

Virtually all coastal groups within the range of the tree considered Douglas-fir wood and bark to be an excellent fuel. The Nuxalk and Stl'atl'imx and the Quinault of Washington made torches from the pitchy heartwood, and the Vancouver Island Salish and the Flathead of Montana used the rotten wood to smoke hides. The Nlaka'pamux mixed fir bark with Ponderosa Pine cones for this purpose.

First Peoples throughout the southern interior used the fragrant boughs of Douglas-fir as flooring and outside covering for sweat-houses; they also used them to cover temporary shelters and ice-fishing holes, as flooring for houses and matting for sitting, for drying food, for butchering deer, to shade fish and berries on drying racks, as scrubbers for initiates and hunters, as padding for packs, and as bedding. They used up to six layers of boughs for bedding, and often spread the fir boughs over layers of other less desirable types of boughs. The Secwepemc used fir branches tied in loose knots and suspended from tree limbs as targets for shooting arrows. Young Secwepemc men tied bundles of the branches to their feet and ran through the water to exercise their leg muscles.

The Okanagan soaked split cedar roots (for weaving) in water containing Douglas-fir needles to impart a yellow tint to the roots. The Swinomish of Washington boiled Douglas-fir bark to make a light brown dye for fishnets, making them invisible to fish. Many groups within the range of the tree used Douglas-fir pitch to seal the joints of implements such as harpoon heads, gaff hooks and fish hooks, and for caulking canoes and water vessels. The Stl'atl'imx mixed it with earth and sand to make soles for fish-skin moccasins.

Western Hemlock	*Tsuga heterophylla*
(Pine Family)	**(Pinaceae)**

Botanical Description
An evergreen tree, Western Hemlock grows 30 to 50 metres tall and about 1 metre in trunk diameter. The crown is narrow, and the top and branches droop, especially in young trees. The bark is thick, dark brown to reddish-brown, and deeply furrowed in older trees. The needles are unequal in length (8 to 20 mm), flat and blunt, with two white bands on the underside, and are usually spread at right angles to the twigs so that the branches are flat. The cones are numerous, about 2 cm long, annual, purplish to green when young and light brown when ripe, opening widely at maturity.

Habitat: humid climates; its shade tolerance is one of the highest among conifers in this region.

Distribution In British Columbia: common along the entire coast of British Columbia to moderate elevations in the mountains where it is replaced by Mountain Hemlock, and recurs in the interior wet belt west of the Rocky Mountains, as far north as the Parsnip River.

Aboriginal Use

Western Hemlock wood is moderately heavy and durable, and works fairly easily. It was used generally along the coast to carve implements such as spoons, roasting spits, dipnet poles, combs, spear shafts, mallets, digging sticks and elderberry-picking hooks. Coastal peoples often made curved fish hooks from the knots formed by the trunk ends of limbs in rotten hemlock and spruce logs. They usually attached barbs made of bone or iron. The Haida Sablefish (Black Cod) hook, also used by neighbouring groups for catching bottom-dwelling fish, is an interesting example of this type of hook. According to Willy Matthews of Massett, the ends of the strongly curved hook were held apart with a small stick, which when a fish was caught, sprung free and floated to the surface indicating the fisherman's success. The Gitxsan used the hemlock knots to make spear points; they and the Nuxalk made wedges from hemlock knots.

The Haida carved large feast dishes out of bent hemlock trunks. They also made fishing weirs, wedges, octopus spears, net anchors, children's bows and ridge-poles for portable houses from hemlock wood. And they sometimes spliced the roots onto Bull Kelp fishing lines to strengthen them. The Nisga'a used hemlock twigs to make rims for their birch-bark baskets.

Western Hemlock bark, which has a high tannin content, was prepared in various ways for use as a tanning agent, pigment and cleaning solution. The Saanich and other Coast Salish peoples pounded and boiled the bark in fresh water to make a reddish dye for colouring Mountain Goat wool and basket materials. Young Saanich women rubbed the dye on their faces as a cosmetic and reportedly to remove facial hair. The Kwakwaka'wakw steeped the bark in urine to make a black dye, and the Nuxalk soaked it in water to make a solution for colouring fishnets brown, making them invisible to fish. They also rubbed the liquid on traps to remove rust and give them a clean scent.

In Washington State, the Clallam, Lummi and Makah pounded the inner bark and boiled it in salt water to make a red paint and wood preservative for spears and paddles. It was said to be more effective when baked over a fire. The Snohomish used the dye to colour basket materials and the Chehalis to tint fishnets. The Quinault mixed the powdered bark with salmon eggs to obtain a yellow-orange paint for staining dipnets and paddles. The Quileute used a hemlock-bark solution for tanning hides and soaking spruce-root baskets to make them watertight.

Western Hemlock covered with herring spawn.

Western Hemlock boughs were considered an excellent bedding material, and were frequently used along the coast for collecting herring spawn. During the spawning season, from March to June, people tied the boughs in bundles and lowered them into the ocean near river estuaries. Later they pulled up the boughs, scraped off the spawn and ate it fresh or dried. The Mainland Comox threaded Eulachon and herring on hemlock boughs for drying, and also used the branches to line steaming pits. The Squamish used them as "rags" to wipe the slime off fish and some people used the boughs to lay fish on for cleaning. Kwakwaka'wakw hunters walking through the forest made trail markers by breaking hemlock branches and turning them back to show the conspicuous white underside. Dancers of Kwakwaka'wakw and other coastal nations wore skirts, headdresses and head-bands of hemlock boughs, and pubescent girls lived in hemlock bough huts for four days after their first menstruation. Many peoples used the boughs as scrubbers for ritual bathing and purification; the Nlaka'pamux name for Western Hemlock means "scrubber plant". Hemlock gum, like that of other conifers, was sometimes used as a glue.

Western Yew
(Yew Family)

Taxus brevifolia
(Taxaceae)

Other Name: Pacific Yew.

Botanical Description
Often small and shrubby, Western Yew grows 5 to 15 metres tall and up to 30 cm in trunk diameter, often twisted and leaning. The bark is reddish, thin and scaly; the wood is strong and flexible. The needles grow alternately from opposite sides of the stem; they are flat and pointed, about 15 to 20 mm long, and brownish-green. The branches superficially resemble hemlock branches but are green beneath, rather than whitish. Male and female reproductive structures grow on different trees. Male trees produce minute yellowish cones, and female trees bear round pinkish-red fruits, consisting of a hard brown seed surrounded by a fleshy cup. These "berries" grow on the underside of the branches and ripen in September.

Habitat: sporadic in moist forests, most commonly found along streams and damp slopes.

Distribution in British Columbia: common on the Pacific coast to the Cascade Mountains, and recurring in the interior wet belt (the Selkirk Mountains and Rocky Mountain Trench).

Aboriginal Use
The heavy, close-grained wood of the Western Yew is well known for its strength and resiliency. It was prized by all First Peoples within the range of the tree, and was frequently traded into areas of the interior where it did not occur naturally. It was used to make implements such as bows, wedges, clubs, paddles, digging sticks, prying sticks, adze handles and harpoon shafts, all of which had to withstand considerable stress. Although tough and hard, Yew wood carves fairly easily and takes a high polish. As an indication of its importance as an implement material, Yew is called "bow plant" or "bow" in a number of First

Nations languages, including Haida, Halkomelem and Stl'-atl'imx, and "wedge plant" in Sechelt, Squamish and Nuu-chah-nulth. Even the Secwepemc in the interior made bows from it, although Yew is scarce in their territory (except in the easternmost

Yew-wood wedge. (RBCM 1995.41.01)

part). The Secwepemc and the Upper Nlaka'pamux often obtained the wood by trade from the Stl'atl'imx and Lower Nlaka'pamux. The Flathead Salish of Montana also made Yew bows, seasoning the wood well and varnishing the finished product with boiled sinew to water-proof it and prevent it from warping. The Saanich of Vancouver Island sometimes moulded the ends of their Yew-wood bows to the proper curvature by steaming them inside a length of Bull Kelp stipe. Some groups, such as the Nisga'a, backed their Yew bows with sinew.

Yew wood was also used to make a variety of other objects, including mat-sewing needles, awls, dipnet frames, halibut and other types of fish hooks, knives, adze handles, dishes, spoons, spears and spear points, boxes, dowels and pegs, drum frames, canoe spreaders, bark scrapers, canoe bailers, fire tongs, combs, and gambling sticks. Harlan Smith (1997) noted that the Gitxsan could buy Yew wood at Fort Simpson. The Stl'atl'imx considered Yew saplings the best material for

snowshoe frames, and recently made Yew-wood handles for shovels and axes. In all cases, people preferred carving the red heartwood over the white sapwood. Aboriginal wood-carvers still like to use Yew, but often have difficulty obtaining large enough pieces. Western Yew is not an abundant species, and though it is little used commercially, habitat destruction from logging has greatly reduced the numbers of large old Western Yew trees in many localities.

The Saanich were said to have used the entire trunk as a catapult in warfare: they would fit a spear to a Western Yew sapling, pull it back and release it. Saanich women used Yew twigs to remove under-arm hair. The Kwakwaka'wakw bound a bundle of Yew branches to a hemlock pole to make a

Alec Peters of Mount Currie (Stl'atl'imx) holding a pair of snowshoes that he has made.

tool for gathering sea urchins; the spines of these animals became entangled in the branches. Young Kwakwaka'wakw men tested their strength by trying to twist a Yew tree from crown to butt. The Quinault of Washington used a Yew trunk as the spring rod in a deer trap, and the Stl'atl'imx used the branches to support deer-hide hammocks. Finally, the Okanagan ground dry Yew wood and mixed it with fish oil to make a red paint.

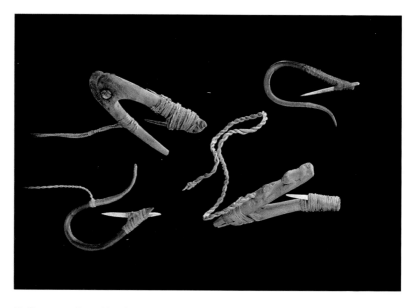

Different styles of halibut and cod hooks. The single-piece curved hooks are made from fir or hemlock knots. (RBCM 711, 714, 2438, 16415)

FLOWERING PLANTS
(Angiospermae)
MONOCOTYLEDONS

Skunk Cabbage
(Arum Family)

Lysichiton americanum
(Araceae)

Other Name: Yellow Arum, Swamp Lantern.

Botanical Description
Skunk Cabbage is a herbaceous
perennial with thick, fleshy rootstocks and large, clustered, oval
leaves, mostly 40 to 100 cm long,
bright green and waxy. Flowers
appear in early spring, consisting
of a bright yellow sheath, up to 20
cm long, surrounding a clublike
yellowish-green flower stalk. At
maturity, the stalk breaks apart to
reveal brown oval seeds embedded in a white pulpy tissue.

Habitat: swampy ground, especially black mucky soil, beneath alder
and conifers; rarely flowers in dense shade.

Distribution in British Columbia: common in coastal forests and
swampy areas from Vancouver Island to Alaska and east to the
Columbia River, but not in dry areas; extends as far north as 54°N latitude in the eastern part of the province.

Aboriginal Use

Skunk Cabbage leaves are not edible, but they are useful in preparing and serving food. Because they are large, flat and water-repellent, some aboriginal people aptly refer to them as "Indian wax paper". They filled the role of wax paper in virtually all coastal aboriginal cultures and even in some cultures of the interior. Skunk Cabbage leaves were widely employed for such tasks as lining steaming pits and covering food being cooked in them, lining and covering berry baskets, lining storage pits (such as for fermented salmon eggs), laying under food, wrapping salmon for cooking, lining oil boxes to prevent leakage, and drying berries and other food on. For drying food, people removed the fleshy mid-ribs and laid the leaves overlapping on the ground or on a wooden rack, sometimes heating the leaves, first, to make them more pliable. Then they put rectangular wooden frames on them and poured the berries – usually cooked to a jamlike consistency – into the frames. Later, when the berry cakes were dry, the leaves could be peeled off the bottom, and the cake was ready for storage. Although Skunk Cabbage leaves have a decidedly acrid odour, they did not seem to impart any unpleasant taste to the foods they came in contact with.

The Nuxalk, Ditidaht and other coastal peoples made an ingenious temporary drinking cup and water dipper by folding a large skunk cabbage leaf in half from top to bottom, bending the two layers to form a U-shaped trough, and pulling the edges back to the lower end, holding them together with the stem as a handle. They also constructed makeshift berry containers by folding and pinning the leaf edges together with sticks. Some people used the larger leaves for sun shades on hot summer days. Children sometimes made throwing spears from the spadix (the club part of the Skunk Cabbage flower) skewered on a stick.

Slough Sedge
Other Sedges
(Sedge Family)

Carex obnupta
Carex species
(Cyperaceae)

Other Names: "Swamp Hay" (all sedges); "Basket Grass", "Swamp Grass", Tall Basket Sedge (*C. obnupta*).

Botanical Description

Sedges are fibrous-rooted, often rhizomatous herbaceous plants that resemble grasses in overall aspect. The stems are usually triangular in cross-section or sometimes rounded, and the leaves are tough, mostly in lines of three with closed (or rarely open) sheaths, with parallel-veined, typically elongated and grasslike blades. Sedges are pollinated by wind. The individual flowers are generally small and inconspicuous, borne on spikes or spikelets with small, brownish scales at the lower end. The male staminate structures and female pistillate structures often grow on different sections of the same spike, on different spikes or even on different plants. The flower spikes are often clustered into compact or open heads, which frequently grow out of elongated leaflike bracts.

Slough Sedge.

Slough Sedge is a relatively large sedge, growing in dense clumps, with long, creeping rhizomes and coarse, stout stems mostly 60 to 150 cm tall, with conspicuous reddish-brown basal membranes. The leaves are coarse and stiff, the blades mostly 3 to 10 mm wide, and more-or-less V-shaped in cross-section. The flower head is surrounded at the base by three bracts (usually). The bracts are sheathless and elongated; the lowest one is usually 10 to 50 cm long and the others progressively shorter. The flowering heads have four to eight cylindrical spikes, mostly 5 to 12 cm long, that droop or spread. The upper one, two or three spikes are staminate, producing pollen, and the others are entirely or partially pistillate. The pistillate scales are narrow, pointed and dark brown with a pale midrib.

Habitat: sedges are most common in moist or wet places, especially in forest swamps, but some species occur in moderately dry to semi-arid sites; Slough Sedge is found in wet meadows and marshes along lake margins, rivers or occasionally in saline coastal swamps.

Distribution in British Columbia: more than 100 species of *Carex* occur throughout the province from sea level to alpine elevations. Slough Sedge grows west of the Coast and Cascade mountains from Haida Gwaii to Vancouver Island and the adjacent mainland; it is one of the most common and widely distributed lowland sedges in the western part of the province.

Aboriginal Use

First Peoples in British Columbia used several kinds of sedges, but it has been difficult to identify them exactly because most aboriginal people do not distinguish the different species with the same general growth habit. Most sedges are simply classed in a general category with grasses and other grasslike plants. Slough Sedge is certainly the most widely used on the coast.

Slough Sedge was, and still is, a popular basket material for the Nuu-chah-nulth on the west coast of Vancouver Island, as well as the Sechelt and other Coast Salish peoples, and the Makah of Washington. The Hesquiat people, north of Tofino, are making a concerted effort to revive and preserve the many facets of their cultural heritage, including weaving with "Swamp Grass". One summer, I had the privilege of helping to prepare some of the leaves for weaving in the traditional man-

Nuu-chah-nulth trinket basket made from Slough Sedge.
(RBCM 13035)

ner. Alice Paul and others in her family had gathered the leaves from swampy meadows north of Hesquiat Village. They gathered only the vegetative or "female" plants. Anyone harvesting the fruiting or "male" plants is laughed at. They pulled up clusters of leaves from the tender white bases, breaking or cutting them off at or just below the ground level. Then they tied them in large bundles and took them back to the village. There, several of us processed the leaves. We removed and discarded the outer leaves, peeled off the inner ones and split them lengthwise exactly in half by running a thumbnail along the midrib. Pieces not split exactly in half were discarded. We then ran each half through the closed thumb and forefinger to flatten it and make it flexible. We had to be careful not to cut our fingers, because the leaf edges are extremely sharp. Finally, we tied the processed halves in bundles of several dozen each and hung them up by their lower ends to dry. Dried bundles are stored for weaving in the fall and winter.

The Hesquiat, Ahousaht and other Nuu-chah-nulth people use a twining process to create the finest baskets and hats from this "grass", often with cedar-bark foundations. They make intricate patterns and designs by weaving in dyed strands of sedge or by superimposing dyed or naturally coloured materials over the regular weave. They weave many styles and sizes of baskets, the most common being round with a flat bottom and fitted lid. After the coming of Europeans it became a widespread practice to weave around bottles and dishes in less traditional forms; synthetic dyes of the brightest hue have almost entirely replaced the soft tones of natural dyes in the designs.

The Squamish, Sechelt, Haida and other coastal groups also used Slough Sedge for weaving, and employed other sedges as well, such as Lyngby's Sedge, a common species of coastal marshes and tidal flats.

The Tule Gatherer (Cowichan) by E.W. Curtis, 1910 (BCARS).

Tule
(Sedge Family)

Scirpus acutus
(Cyperaceae)

Other Names: Hard-stemmed Bulrush, Roundstem Bulrush, Bulrush, Rush.

Botanical Description

Tule is a stout, rhizomatous perennial, usually 1 to 3 metres tall, that often grows in wetlands in dense colonies. The stems are round, swollen towards the lower end to a diameter of 2 cm or more and gradually tapering towards the top; they are pithy but tough. The stem bases are white and succulent, the tops dark green. The leaves, borne at the base of the stem, consist of two or three prominent brownish sheaths with or without short, poorly developed blades. Alongside the flower head is an erect, tapering green bract 2 to 10 cm long, appearing as a continuation of the stem. The flowers grow in numerous compact

grey-brown spikelets clustered at the ends of a number of short branches spreading from a single point at the top of the stem. *Scirpus validus* (also called Tule, or Soft-stemmed Bulrush), classed by some botanists as a separate species, is included in this discussion because First Peoples consider it the same as *S. acutus.* Some botanists include both *S. acutus* and *S. validus* as subspecies of *S. lacustris.*

Habitat: marshes and swampy ground at the edges of lakes and streams at lower elevations; sometimes growing in water a metre or more deep.

Distribution in British Columbia: widespread in the province in appropriate habitats, especially in the central and southern interior, where it often forms extensive colonies around alkali lakes.

Aboriginal Use

Tule was – and still is – an important mat-making material for many of the province's aboriginal peoples, especially the Coast and Interior Salish. The Nuu-chah-nulth, Kwakwaka'wakw, Carrier and Ktunaxa also used it. Most people harvested the tall, round stems at the peak of maturity in late summer and early fall, but sometimes they gathered them in late November after the stems turned brown. If the plants are too young they are virtually impossible to pull up, but at the right stage they break off easily at the base with only a slight tug on the upper stem. If the water is not too deep where they are growing they can be cut with a

Coast Salish mats, Tule on top and Cattail beneath.

knife. Care must be taken not to bend or kink the stems during harvesting.

Tule harvesters tied the stems in large bundles to keep them from breaking, then carried them home where they spread them out and dried them in the sun. To make a mat, they laid Tule stems side by side, alternating top and bottom, and either sewed them with a long wooden needle or twined them together with a tough fibre such as Stinging Nettle, Indian Hemp or, within the last century, cotton string. Tule mats are light, with a good insulating capacity because of the stems' pithy centres, and they can be rolled easily longitudinally into a tight bundle. They served many purposes. The largest were used for the roofs and walls of temporary shelters, summer dwellings and tipis, and as insulation for the walls of winter houses. Medium-sized mats were used as door covers, rugs, mattresses and wind breaks, for drying berries, and for cutting and drying meat and fish on. The smaller ones were used for covering windows, for sitting on at home or in the canoe, and for eating on. Mattresses and sitting mats could be made extra soft by piling two or more on top of each other.

The Okanagan made large bags from Tule twined with various other fibres, including Silverberry bark, willow bark or Indian Hemp, using them to store dried roots, berries and fish. The Okanagan used to play a ball and pin game with a small ball made out of Tule. They made the ball by folding a Tule stem back and forth over itself and tying it; the pin was from a Red Hawthorn spine. They also made Tule headdresses for aboriginal doctors. The Nuu-chah-nulth made baskets, basket lids and, recently, handles for shopping-bags from Tule. The Vancouver Island Salish traded Tule mats to the mainland Salish in exchange for

American Bulrush.

Mountain Goat wool and wool blankets, and in the 19th century, exchanged them with the Nuu-chah-nulth for halibut.

The First Peoples of this region used two other species of *Scirpus* in their technology: American Bulrush is a tall, triangular-stemmed species, and Small-flowered Bulrush, which is also commonly known as "Cut-grass" because of its broad, razor-sharp leaves. Both, like Tule, grow in marshes. The Nuu-chah-nulth and others used American Bulrush – which some people call "Three Square" and some call "Sweetgrass" – as the foundation for their tightly twined trinket baskets. The weaving strands for these are Slough Sedge, Bear-grass and, recently, Raphia leaves. Haisla children sometimes made snares for small fish from American Bulrush. Haisla people also made basket lids and handles from the stems. The Okanagan wove berry and root baskets from the stems of Small-flowered Bulrush and possibly other types of sedges. They also wove the dried leaves into buckskin dresses as trimming and sometimes laid them over and under food in steaming pits. The sharp-edged leaves of "Cut-grass" could be used as cutting implements, but they were usually too fragile to be of much use.

A Nuu-chah-nulth trinket basket made with a foundation (warp) of American Bulrush and wefts of Slough Sedge and Raphia.

Bear-grass
(Lily Family)

Xerophyllum tenax
(Liliaceae)

Other Names: Deer-grass, Basket-grass, Squaw-grass, Pine Lily, "American Grass".

Botanical Description
Bear-grass is a large perennial herb with a short, stout rhizome and numerous tough, pointed grasslike leaves. Usually 15 to 60 cm long, the leaves form a dense clump at the base of an erect, leafy stem that often exceeds a metre in height. The flowers are numerous, small, white and fragrant, and they grow in a dense cluster, spherical but pointed, at the end of the stem.

Habitat: open woods and meadows usually at higher elevations.

Distribution in British Columbia: restricted to the southeastern corner of the province, but occurring near sea level on the Olympic Peninsula in Washington and eastward to the Rocky Mountains.

Aboriginal Use
The First Peoples of British Columbia and Washington used the tough, minutely serrated leaves of Bear-grass in basketry, especially for fine imbrication, trimming and ornamentation. The Nuu-chah-nulth and Coast Salish of Vancouver Island used this grass, as did the Squamish, Sechelt, Nlaka'pamux, Okanagan and Ktunaxa. Of these groups only the Ktunaxa had direct access to the plant; the others obtained it through trade from neighbouring Washington groups or from the Ktunaxa. Bear-grass leaves were a common item of commerce at the Columbia Rapids, the centre of the Chinook Salmon trade. In the mid 20th century, a small bundle of prepared Bear-grass leaves cost about 50 cents.

After harvesting the leaves, people cut them to a uniform width using a gauged knife edge. Bear-grass dries to a lustrous, creamy-white colour. It was often used in its natural colour, but also takes dyes well, either vegetable pigments or (unfortunately in the opinion of some) synthetic dyes. As early as 1902, Dr C.F. Newcombe was prompted to remark, "The fatal facility with which *Xerophyllum* takes aniline dyes ... and the demands of the average collector for gaudy shades has quite demoralized the colour sense of the Nootkans [Nuu-chah-nulth] and their recent basket work is particularly discordant." Nevertheless, many people appreciate the bright colours of the 20th century baskets, and weavers have experimented with new colours, such as those made from Kool-Aid crystals and shoe dye, for their basketry materials.

As well as for trimming and imbrication, the Nuu-chah-nulth used Bear-grass leaves for the weft in their finely twined baskets and hats (see the photograph on page 125), and also wove them into mats. The small baskets they sold to the Hudson's Bay Company for blue trading beads. They obtained their Bear-grass leaves, which they called "American Grass", from the Makah and Quileute peoples of the Olympic Peninsula. The Ktunaxa used the leaves for hats. The Southern Okanagan and Sanpoil-Nespelem Okanagan of Washington, who obtained the leaves from the neighbouring Kalispel and Pend d'Oreille territories, used them to make designs on birch-bark baskets. Recently, people have substituted corn husks when they could not obtain Bear-grass.

Bluebunch Wheatgrass.

Grasses   (Poaceae or Gramineae)

British Columbia's First Peoples used grasses for a multitude of household tasks: to line steaming pits, wipe fish, cover berries and spread on floors, to string clams, Eulachon and roots for drying, and as bedding. Coast Salish hunters lured deer by whistling through a blade of grass. The Kwakwaka'wakw used fresh green grass to colour abalone shells and spruce-root hats, and the Nuu-chah-nulth made a green dye for painting wood and basket materials by macerating grass, soaking it for a long time in hot water, and drying it into a cake, which was pulverized and mixed with oil.

A number of grasses – including Pinegrass, Giant Wildrye, Common Sweetgrass, Reed Canary Grass and Common Reed Grass – were fairly important in aboriginal technologies and are discussed in detail on the following pages. Several others were used in minor ways by the First Peoples of British Columbia and their neighbours. The Okanagan used Bluebunch Wheatgrass (also called Bunchgrass), a dominant species of the dry interior, as tinder for starting fires and for stuffing moccasins in winter. They also inserted Bluebunch Wheatgrass straws into newly pierced ears to keep the openings from sealing. The Stl'atl'imx dried Saskatoon berries on Bluebunch Wheatgrass and sometimes used it to make a Soapberry beater. They also said that it is a favourite forage crop for deer and makes excellent hay for livestock. The Secwepemc and other interior peoples used Bluebunch Wheatgrass for lining their root-cooking pits. The Gitxsan used grass for stuffing moccasins and babies' bedding, and for sitting on. Okanagan children played with the awned seeds of "Speargrass" (Needle-and-thread Grass), throwing them like darts. The Blackfoot of Alberta bound the awns of Porcupine-grass into a cylindrical bundle the size of a man's thumb and used this as a hairbrush. The Tlingit of Alaska used the split stems of several species of grasses to imbricate their fine spruce-root baskets, including brome grasses, Bluejoint, Nodding Woodreed, Tufted Hairgrass and Fowl Mannagrass. The Dena'ina of Alaska burned grass as a smudge against mosquitoes, and used grass in a variety of other ways: cutting fish on, marking trails over glaciers, filling in swampy places on a trail, covering floors, thatching roofs, insulating footwear and lining cooking pots.

Pinegrass
(Grass Family)

Calamagrostis rubescens
(Poaceae)

Other Name: "Timbergrass".

Botanical Description

Pinegrass is a perennial with creeping rhizomes and slender, tufted stems 60 to 100 cm tall and usually reddish at the base. The leaves are erect, flat or V-shaped, and 2 to 4 mm wide, with smooth but slightly hairy sheaths. The flowers, one per spikelet, grow in a narrow, greenish to purplish, irregular cluster 7 to 15 cm long. In many places, the flowers are not common and the plants spread vegetatively.

Habitat: a valuable range grass of open, dry sagebrush flats to moist montane forests, often forming extensive patches that cover the forest floor.

Distribution in British Columbia: abundant in the southern interior from the Chilcotin to the lower Columbia River valley.

Aboriginal Use

Interior peoples used Pinegrass, widely known as "Timbergrass", in a variety of ways. They gathered Pinegrass leaves in the summer by the handful. They cut off the root ends and used the leaves to whip up Indian ice cream, a frothy confection made from Soapberries. Sometimes they tied a bundle of grass onto a stick to make a Soapberry beater. A common practice among the Nlaka'pamux, Secwepemc and others was to spread cooked Soapberries on a mat of Pinegrass leaves that were loosely braided together at the ends and laid out on a rack. After drying the berries over a small fire, they stored them for winter with the grass still attached. To make the Indian ice cream, they soaked pieces of the dried cakes in water and whipped them into foam with their hands; the grass helped to whip the berries. Then they skimmed the grass leaves off the top of the Soapberry whip.

The Okanagan used Pinegrass to line and cover berry baskets, and to place over and under the food in steaming pits. To help keep their feet warm in winter, they wove socks and moccasin insoles from Pinegrass leaves, first rubbing them together to soften them. They also twisted the leaves into twine and mixed them with mud to chink log cabins. Stl'atl'imx hunters washed their guns and traps in water using Pinegrass as a sponge.

Giant Wildrye (Grass Family)

Elymus cinereus (Poaceae)

Other Name: Giant Ryegrass.

Botanical Description
Giant Wildrye is a robust perennial that forms large clumps with stout, erect stems, often 1 to 2 metres high. The leaf blades are firm, flat, strongly nerved, and up to 1.5 cm wide, the sheaths smooth to densely hairy. The flowers, two to six per spikelet, grow in a dense wheatlike spike 10 to 25 cm long. The spikelets are usually in clusters of three to five per node.

Habitat: river banks, gullies, dry washes, moist or dry slopes and plains, in sandy or gravely soil.

Distribution in British Columbia: common in the dry interior, from the Thompson River to the Rocky Mountain Trench.

Aboriginal Use
Interior Salish peoples, especially the Nlaka'pamux, imbricated split cedar-root baskets with Giant Wildrye. They cut the stiff, hollow fruiting stems while still green and "smoked" them over a hot fire to prevent them from turning brown, then they split and washed them. Some Okanagan groups made small, temporary arrows from the stems, notching them and fixing them with tips of Mock-orange wood. They used these arrows in games and for hunting small birds. Similarly,

young Flathead boys fixed hawthorn points to the stems and used them as spears to inflict pain upon each other in preparation for warfare. The Okanagan lined steaming pits and food caches with the leaves; they also covered the floors of winter houses and sweat-houses with them if Douglas-fir boughs were not available, and used them as bedding and horse fodder. They set a hollow Giant Wildrye stem into the centre of a food cache before it was covered over to prevent the food from "sweating" and becoming affected by mildew. The Stl'atl'imx used the leaves, with Saskatoon Berry branches, for lining steaming pits, and the Blackfoot of Alberta used them for bedding. A closely related species, Creeping Wildrye, was reported by Dr C.F. Newcombe (in his unpublished notes, ca 1903) and James Teit (in Steedman 1930) to be used by the Nlaka'pamux and Stl'atl'imx to make baskets and to trade to the Sto:lo, Squamish and Sechelt on the coast. But in British Columbia, Creeping Wildrye grows only in an isolated location on the Saanich Peninsula on Vancouver Island; the species used must have been a related one, possibly Blue Wildrye or Canada Wildrye. In any case, the stems were the part of the plant used, being cut, slit down one side, flattened and superimposed over the weave of baskets.

The Vancouver Island Salish used the leaves of another species, Dunegrass, to weave tumplines and pack straps and for tucking into the ravels of reef nets to strengthen them. The Nuu-chah-nulth sometimes used Dunegrass leaves for weaving basket handles, and the Quinault of Washington wove tumplines from them and spread them out to dry Salal berries on. The Haida split the stems in half, dyed them, and used them to twine and decorate baskets. The Haisla used this and other grasses to line Eulachon-ripening pits and for tying bundles of Silverweed roots together for steaming. The Tlingit also used the stems as overlay for baskets, but only for coarse work or when the other grasses, mentioned on page 116, were not available. The Dena'ina of Alaska used Dunegrass to weave mats for sitting on, and also used it, dyed or undyed, in their coiled baskets.

Dunegrass.

Common Sweetgrass
(Grass Family)

Hierochloë odorata
(Poaceae)

Other Names: Holy-grass, Vanilla-grass, Seneca-grass.

Botanical Description

Common Sweetgrass is a reddish-based perennial with slender, creeping rhizomes and leafy stems 30 to 50 cm tall. The leaf blades are usually 2 to 5 mm wide and pointed. Those on the stem are fairly short, while those on vegetative shoots are up to 25 cm long. The flowers are small, three per spikelet, in an open pyramidal cluster. The leaves have a sweet, vanilla-like fragrance. In various parts of North America, several grasslike plants are locally called "Sweetgrass" – one of these is American Bulrush, a relative of Tule.

Habitat: moist meadows and slopes from moderate to subalpine elevations, ranging down to sea level in some places.

Distribution in British Columbia: widespread in the province, but seldom abundant.

Aboriginal Use

The sweet, lingering fragrance of Common Sweetgrass is due to the presence of coumarin, a fragrant crystalline compound that was once used commercially as a flavouring. First Peoples throughout North American appreciated Common Sweetgrass for its scent. In the East, the Seneca and others wove fragrant baskets from the leaves. The Flathead of Montana and the Blackfoot of Alberta plaited bundles of the dried leaves into a thick, three-ply braid, which was used as a sachet in clothing or ignited at one end and burned as an incense, air and clothing freshener, and insect repellent. They also placed the leaves on a hot stove and allowed them to smoke. Blackfoot women often wore a plaited band of Common Sweetgrass around their heads, braided it into clothing, or carried it around in a small buckskin bag as a perfume.

In British Columbia, the Ktunaxa used Common Sweetgrass in a similar manner. They often purchased the thick braids of leaves from the Blackfoot and burned them like a punk to fumigate a house. They say that up in the mountains in some localities one can smell the fragrance of Common Sweetgrass on the wind. The Nlaka'pamux also used it, crumpled up in a bag as a sachet, tied in the hair or on neck and arm ornaments, or rubbed on the clothing, hair and skin.

Common Sweetgrass also grows in areas of coastal British Columbia, such as the Kitlope River valley where it was apparently used by some Haisla women to make baskets. They gathered the grass in May and June when it was about 30 cm tall. The identity of this grass, as reported by Compton (1993), is still tentative.

Margaret Lester and Nellie Peters of Mount Currie harvesting Reed Canary Grass.

Nellie Peters showing how bundles of grass are dried. The dried grass stems are to be used to decorate coiled cedar-root baskets.

Reed Canary Grass
(Grass Family)

Phalaris arundinacea
(Poaceae)

Botanical Description
Reed Canary Grass is a tall, coarse perennial grass with creeping rootstocks and stout, erect stems up to 1.5 metres high or more. The leaves are flat and numerous, up to 12 mm wide. The flowers grow in dense, compound clusters up to 18 cm long; they are narrow and erect when young, and spread slightly at maturity. Reed Canary Grass is an important hay species, but said to be coarse and not as nutritious as some other grasses.

Habitat: swamps, lake margins, roadside ditches and moist meadows, often in standing water.

Distribution in British Columbia: common throughout southern British Columbia and north to the Peace River.

Aboriginal Use
The Upper Sto:lo of the Fraser River, the Lower Stl'atl'imx and probably other Salish groups imbricated coiled cedar-root baskets with the stout, smooth stems of Reed Canary Grass. They gathered pliable, green stems in May and early June, around the time when wild roses bloom, cut them into even lengths and soaked them in boiling water, then dried them in the sun for several days to bleach them white. They split the dried stems, soaked them, and used them, like those of Common Reed Grass (see the following account), to superimpose white patterns on the weave of split-root baskets. Some Stl'atl'imx pack baskets are completely white over more than half their surface from these grass stems. Contrasting patterns are made from natural-red and dyed-black Bitter Cherry bark. The Okanagan used Reed Canary Grass to make eating and food-drying mats, for weaving peaked hats for their doctors, and for binding fishing weirs.

Common Reed Grass
(Grass Family)

Phragmites australis
(Poaceae)

Botanical Description

Common Reed Grass is a tall perennial with stout, creeping rhizomes and erect, leafy, conspicuously noded stems 2 to 4 metres tall. The leaf blades are flat, 20 to 40 cm long and up to 5 cm wide; the sheaths are

smooth and loose, often twisting in the wind so that the leaves are all aligned in the same direction. The flowers grow in dense terminal clusters, 15 to 35 cm long; they are purplish to white, and fuzzy, resembling small heads of the ornamental pampas-grass. *Phragmites australis* is also known as *P. communis*.

Habitat: marshes, lake and river margins, and roadside ditches, often in standing water.

Distribution in British Columbia: in the southern half of the province, from Vancouver Island to the Okanagan, often forming dense colonies.

Aboriginal Use

The Nlaka'pamux, Stl'atl'imx and Carrier, and probably other groups as well, used the smooth, glossy, hollow stems of Common Reed Grass to imbricate split cedar-root baskets, in much the same way as they used Giant Wildrye stems. James Teit considered Common Reed Grass "one of the most commonly used basketry materials" of the Nlaka'pamux (Steedman 1930). To prepare the grass, people split the stems down one side, then flattened and dried them. They usually left the stems in their natural creamy white colour, but sometimes dyed them yellow or some other colour. In creating decorations, basket makers interspersed Common Reed Grass stems with natural-red and dyed-black strips of Bitter Cherry bark.

The Nlaka'pamux made mats from lengths of Common Reed Grass stems. They also dyed the stems and used them with various seeds and berries as decorative beads on necklaces and fringes of dresses. The

Blackfoot of Alberta used the stems for arrow shafts and pipe stems. The Secwepemc sometimes used Common Reed Grass hay as bedding and for spreading around a camp to keep the dust down.

Nlaka'pamux elder Mabel Joe of Nicola Valley displaying Common Reed Grass berry-drying mats she has made.

Cattail
(Cattail Family)

Typha latifolia
(Typhaceae)

Other Names: Common Cattail, Bulrush (this name also applies to several species of *Scirpus*, in the sedge family).

Botanical Description
Cattail is a tall perennial with thick, white, fleshy rhizomes and long swordlike leaves that are greyish-green and usually 1 to 2 cm wide. The flowers are contained in a compact, brown spike, familiar to almost everyone as the cat's tail. Pollen is produced on a thinner spike immediately above the brown portion. The mature fruits are released in late summer and fall, and the cat's tail becomes a mass of whitish fuzz that is gradually blown away in the wind.

Habitat: shallow marshes and swamps, and lake edges, often forming extensive pure stands.

Distribution in British Columbia: throughout the province, except on the central and northern coast, and only recently introduced to Haida Gwaii; most common around ponds and lakes in the southern part of the province.

Aboriginal Use

The flat, pithy leaves of Cattail were, along with Tule stems, the most important mat material of the Salish peoples in the province, and were used by other groups as well. Most people gathered the leaves in late summer, cut them to even lengths and dried them in the sun. They constructed mats by laying the leaves out side by side, alternating top and bottom, and threading them together transversely at about 10-cm intervals, using a plant fibre such as nettle twine, or the lower edge of the Cattail leaf itself. For this procedure, they used a long, thin needle of "Ironwood" (Oceanspray) or some other hardwood. The mat maker poked the needle through an entire row of leaves and firmly pressed the leaf tissue around it with a grooved mat creaser (see page 126), often made of Broad-leaved Maple wood, to make an opening for the thread and to crimp the leaves so they would not spread. To prevent the mat from unravelling, the mat maker sewed the braided leaves at the edges, folded the ends over and bound them. Cattail mats could be two metres long and nearly as wide. Cattail leaves could also be twined rather than sewn together to make mats. The Coast and Interior Salish, Nuu-chah-nulth, Kwakwaka'wakw and Ktunaxa used Cattail mats to make the walls and roofs of summer houses, insulate the walls of winter houses, kneel on in canoes, sit on, dry berries on, and cover doors and windows; they also used them as saddle blankets, mattresses or mattress underlays, and carpeting. For extra softness, they piled three or four mats on top of each other. The Nuu-chah-nulth and Kwakwaka'wakw sometimes obtained Cattail leaves or finished mats from the Salish through trade. Aboriginal elders generally maintain that these peoples did not make Cattail mats originally, but learned the art from their Salish neighbours within the last century.

Several groups also used Cattail leaves to make twine, baskets, bags, capes, hats, and doctors' headdresses. The Saanich split the leaves and spun them on their bare thighs to make storage baskets for Blue Camas bulbs and crabapples. They also made a baby's first cradle from bundles of Cattail leaves tied together. The Lower Stl'atl'imx of the Pemberton area made strong, four-, six-, and eight-strand ropes by plaiting the leaves with the bark of Red-cedar roots. Squamish and other Coast Salish peoples made a tough fine thread by stripping off the thin edges of the whitish leaf bases with the fingernail, drying them, and rolling them together on the thigh.

The Straits Salish used charcoal from burned Cattail leaves for tattooing, and the Kwakwaka'wakw mixed the charcoal from old burned Cattail mats with water and dried herring spawn and used it as a paint for the insides of canoes to protect them from weathering.

Cattail seed fluff was used by the Interior Salish to stuff pillows and mattresses. One old-timer of the Lower Stl'atl'imx people remembers his mother collecting it by the sackful for this purpose. The Okanagan, Secwepemc and other peoples used it for dressing wounds and as babies' diapers. The baby's urine caused the fuzz to form little balls, which were absorbent and easy to remove from the cradle. The Saanich sometimes spun the "cotton" seed fluff with dog wool to make blankets.

Fishing Camp – Skokomish by E.W. Curtis. The temporary shelter is made of Cattail mats. The people are wearing clothes made of cedar bark and furs. In the foreground is a dugout cedar canoe. (RBCM PN6449)

Common Eel-grass	*Zostera marina*
Scouler's Surf-grass	*Phyllospadix scouleri*
Torrey's Surf-grass	*Phyllospadix torreyi*
(Eel-grass Family)	(Zosteraceae)

Other Names: Ribbon-grass (*Z. marina*); Sea-grass (*P. torreyi*); Basket-grass (*Phyllospadix* spp.).

Botanical Description
Common Eel-grass, Scouler's Surf-grass and Torrey's Surf-grass are the three main marine species of flowering plants on British Columbia's coast. They are perennials with long, flexible ribbonlike leaves and fleshy rhizomes. Common Eel-grass has flat bright-green leaves that are more than 32 mm wide; two varieties occur: *Z. marina* var. *marina,* with leaves usually less than 1.2 metres long, and *Z. marina* var. *latifolia,* with

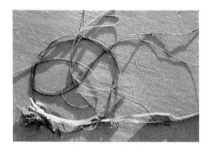

Scouler's Surf-grass (above) and Common Eel-grass (below).

leaves up to 4 metres long. The two surf-grasses have bright emerald-green leaves that are narrower than those of Common Eel-grass; Scouler's Surf-grass leaves are flat and usually under a metre long, and Torrey's Surf-grass leaves are rounded and up to 3 metres long. The flowers of all three species grow inconspicuously at the bases of the leaves; they are pollinated under water with the aid of ocean currents.

Habitat: Common Eel-grass occurs in marine bays in mud or sand in the intertidal zone; both surf-grasses usually occur on rocky wave-swept shores.

Distribution in British Columbia: all three species are common along the British Columbia coast and are often seen washed up on beaches after a storm.

Aboriginal Use

The Vancouver Island Salish and Kwakwaka'wakw used the damp leaves of Common Eel-grass to generate steam in the board-bending process of making kerfed cedar-wood boxes and in other types of wood moulding. The Haida, Nuu-chah-nulth and other coastal groups occasionally collected herring spawn from the leaves of Common Eel-grass and both surf-grasses. The Nuu-chah-nulth and Makah used the leaves of all three species, especially those of Scouler's Surf-grass, to imbricate baskets and hats. They sun-bleached the leaves to a bright white, or dyed them any colour they required for a basket pattern. Quileute boys of Washington sometimes used bunches of the leaves as targets in arrow practice, but according to Erna Gunther (1945), the Quileute never used surf-grasses in basketry. Evidently, the Kwakwaka'wakw wove belts and baskets from Common Eel-grass, but these would not have been very strong.

Nuu-chah-nulth whaler's hat woven by Ellen Curley of Clayoquot. The hat is made of Common Eel-grass and Bear-grass on cedar bark. The whaling design is made with Eel-grass, some bleached, some coloured with mud and some dyed with Oregon-grapes. (RBCM 9736)

Cowichan Cattail mat creaser made of maple wood. (RBCM 10695) See page 122 for a description of how it was used.

FLOWERING PLANTS
(Angiospermae)
DICOTYLEDONS

Vine Maple
(Maple Family)

Acer circinatum
(Aceraceae)

Botanical Description

Vine Maple is a small tree or multi-stemmed shrub 3 to 10 metres tall, with branches in opposite pairs. The bark is smooth and pale green to purplish-red. The leaves, 5 to 13 cm across, are palmately veined and have seven to nine pointed lobes with saw-toothed margins; the leaves are bright green, turning orange to  scarlet in autumn. The small flowers have purplish sepals and white petals, and grow in loose, clusters of only a few flowers. The fruits, growing in pairs, are smooth with prominent, widely spreading wings. Rocky Mountain Maple (see next account) is sometimes locally called Vine Maple, which can cause confusion in identification.

Habitat: moist, shaded woods from sea level to subalpine elevations.

Distribution in British Columbia: on the southern coast west of the coastal mountains, but with a few isolated reports farther inland; rare on Vancouver Island and restricted to just a few locations there.

Aboriginal Use

Vine Maple wood is hard, but limited in size and inclined to warp with time. The Quinault of Washington made fish traps and large, loosely woven carrying baskets from Vine Maple splints. The Skagit of Washington made salmon tongs from the wood, the Katzie (Sto:lo) of the Fraser River valley used it for spoons, and the Squamish and Cowichan for knitting needles. The Lower Stl'atl'imx and Lower Nlaka'pamux made snowshoes, slat-armour vests, arrows, baby-basket frames, implement handles and, sometimes, dipnet frames from Vine Maple, as well as from Rocky Mountain Maple. They, the Squamish and Katzie sometimes made bows from the long, straight branches of Vine Maple. The Squamish also made dipnet frames from it. All these groups sometimes used the wood for fuel. The Quinault made black paint from the charcoal mixed with oil.

Rocky Mountain Maple (Maple Family)

Acer glabrum (Aceraceae)

Other Names: Mountain Maple, Douglas Maple; sometimes mistakenly called Vine Maple.

Botanical Description

Rocky Mountain Maple is a bushy shrub or small tree, 1 to 10 metres tall, with opposite branches and smooth, greyish to reddish-purple bark. The leaves are 6 to 10 cm long and about the same across. They are palmately veined, with three to five sharp, coarsely toothed lobes;

they turn a bright red-orange in the fall. The tiny yellowish flowers grow in small, loose clusters. The fruits are smooth and grow in pairs; they have prominent wings, usually spreading at less than a 90° angle.

Habitat: coastal lowlands to dry rocky slopes, usually in open locations.

Distribution in British Columbia: from the southern coast of Vancouver Island to the Rocky Mountains and northward to Alaska and Dawson Creek; abundant and widespread in the southern interior; not found on Haida Gwaii.

Aboriginal Use

Many aboriginal people, especially in the interior, refer to Rocky Mountain Maple as Vine Maple; in areas where its range overlaps with that of the true Vine Maple, it is often difficult to tell which species is being referred to. The tough, pliable wood of Rocky Mountain Maple was employed in many ways. The Nlaka'pamux, Okanagan, Secwepemc, Stl'atl'imx, Gitxsan, Haisla and Carrier commonly used it to make snowshoe frames. The Haisla called it "snowshoe tree". These groups also used Rocky Mountain Maple to make other items: the Nlaka'pamux for bows, baby swings, and the hoods of baby cradles and baskets; the Okanagan for drum hoops, tipi pegs and tongs; the Secwepemc for digging sticks, fish traps, scoop-net handles, spear prongs, and the shafts of spears and harpoons; the Stl'atl'imx for bows, arrows and combs; the Gitxsan for spoons, rattles, paddles and arrows; the Haisla for spoons, axe handles and baskets; and the Carrier for labrets. The Oweekeno called Rocky Mountain Maple the "spoon tree" and the Nuxalk also used it for making spoons. The Sekani made bows from Rocky Mountain Maple wood; the Tsilhqot'in made throwing sticks from it; the Nuxalk made spoons and slat armour; the Nisga'a made raven rattles, masks and headdresses; and the Haida, who acquired it by trade from the Tsimshian on the mainland, made grease dishes, Soapberry spoons, Sea Otter clubs, dipnet handles, gambling sticks and totem models. The Flathead of Montana used it to make arrow shafts, pipe stems and sweat-house frames. The green wood could be easily moulded by first soaking it in water, then heating it over an open fire and bending it to the desired shape while still hot.

Rocky Mountain Maple wood was considered an excellent fuel. The Ktunaxa used it as the drill

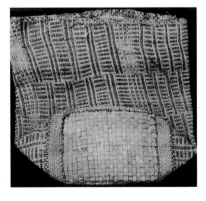

Nisga'a bag woven with natural and black-dyed Rocky Mountain Maple bark fibres. (RBCM 1684)

and hearth in making friction fires. In the old days, the Nuxalk sometimes felled Red-cedar trees with a long maple branch. They bound the
branch around the base of a tree and ignited it. Once it caught fire, the
branch smouldered for a long time, eventually burning right through
the tree. Near Bella Coola village, a cedar tree, partially burned through
by this method many years ago was still standing as of 1980, with a
maple branch embedded in the wood.

The Stl'atl'imx, Secwepemc and others used the fibrous bark of this
maple to make rope, to string roots (such as Yellow Avalanche Lily
bulbs) together for drying, and tied in a bundle to a stick, to whip
Soapberries. The Gitxsan and Nisga'a wove mats, baskets and pack
sacks from Rocky Mountain Maple bark, using black- and red-dyed
strands to make decorative patterns. The Blackfoot of Alberta made
paint cases from folded sheets of the bark. The Clayoquot Nuu-chah-
nulth peoples made loosely woven baskets of "Vine Maple" (evidently
Rocky Mountain Maple) and the Ditidaht made drinking bowls and oil
dishes from either this or the true Vine Maple. Gitxsan children shook
clusters of the dried seeds as toy rattles.

Broad-leaved Maple
(Maple Family)

Acer macrophyllum
(Aceraceae)

Other Names: Big-leaf Maple, Common Maple.

Botanical Description
Broad-leaved Maple is a large, spreading tree, up to 30 metres tall. The
trunk and branches are often moss-covered. The bark on young

branches is smooth and greenish
or reddish; on the older branches
and the trunk it is grey and furrowed. The leaves are the largest
of any tree in the province, up to
30 cm or more wide. They are
smooth-edged and have five
prominent, pointed lobes; the
larger ones also have smaller lateral lobes. The flowers are small

and yellow-green, and hang in long clusters. The seeds, usually in pairs, are hairy and have prominent wings.

Habitat: damp woods and slopes at low elevations.

Distribution in British Columbia: along the southern coast and sporadic northward to Alaska, west of the coastal mountains, and on Vancouver Island; not found on Haida Gwaii.

Aboriginal Use

The Coast Salish people and, to a lesser extent, the Nuu-chah-nulth and Kwakwaka'wakw, carved spindle whorls and paddles from Broad-leaved Maple wood. In fact, the name for this maple in a number of Coast Salish languages means "paddle-tree". From it these peoples also carved dishes, spoons, fishnet measures, fish lures, hairpins, combs, balls, Cattail mat creasers, rattles, cedar-bark shredders and adze handles. The Lower Stl'atl'imx of the Pemberton area sometimes used the wood for pipe stems and snowshoe frames, although they usually preferred Rocky Mountain Maple for the latter. Most people considered Broad-leaved Maple an excellent fuel: it burns with a hot, smokeless flame. The Squamish, and the Swinomish, Chehalis and Quinault of Washington smoked fish with the decayed wood.

The Kwakwaka'wakw wove open-work baskets and fish traps, and the Cowlitz in Washington made ropes and tumplines from the inner bark of Broad-leaved Maple. The Lower Nlaka'pamux wove bark fans and sometimes tied a bundle on a stick, like Rocky Mountain Maple bark, to make a small whisklike beater for whipping Soapberries.

Some peoples, such as the Straits Salish and Squamish, occasionally whipped Soapberries with the large Broad-leaved Maple leaves tied in a bunch. Many groups spread the leaves under and over food in steaming pits to impart a pleasant flavour to cooking meat. They also used leaves to line berry baskets and even make small, crude berry containers. The Comox used them to line the pits in which

Coast Salish spindle whorl carved from Broad-leaved Maple. The faces in the centre are surrounded by four Minks. (RBCM 9658)

salmon roe was buried to make "stink eggs" (fermented salmon eggs), a great delicacy for some. The Squamish used them as rags to wipe the slime off freshly caught fish.

Cow Parsnip   *Heracleum lanatum*
(Celery Family)   (Apiaceae)

Other Names: "Indian Rhubarb", "Wild Rhubarb".

Botanical Description
A robust, hollow-stemmed perennial, Cow Parsnip grows 1 to 2 metres tall from a stout tap root or root cluster. The leaves are broad and compound in three large segments (one terminal and two lateral), coarsely toothed and lobed. The flowers are small, white and numerous, arranged in large flat-topped umbrella-like clusters. The leaf stems are conspicuously winged at the base. The plants have a pungent odour, especially when mature.

Habitat: moist open areas, roadsides and meadows, from sea level to above the timberline in the mountains, often in large patches.

Distribution in British Columbia: throughout the province; common in the north. The Parsnip River, a tributary of the Peace, is named after Cow Parsnip.

Aboriginal Use
The Secwepemc covered berry baskets with the large leaves of Cow Parsnip, and the Carrier dried Saskatoon berries on them. The Secwepemc and the Flathead of Montana used the dried hollow stems to make Elk and Moose whistles, while the Haida used them as moulds for making spruce gum "dice", employed in a game. The Blackfoot of Alberta made children's flutes, drinking straws and toy blowguns from them. Washington Makah and Quileute girls entwined the large umbrella-spoked flower heads of Cow Parsnip with seaweed to make play-baskets for holding shells and small objects. The Secwepemc boiled the entire plant in water to make a washing solution for eliminating fleas from clothing.

The Gitxsan and Haisla used the stems for whistles, and Gitxsan girls used them as drinking straws during puberty rites. The Dena'ina of Alaska also used the stems as straws. Children in many parts of the province made blowguns from hollowed elderberry stems or Cow Parsnip stems, using pieces of kelp or other vegetation as ammunition (see the warning below.)

Warning

Blowguns or straws made from Cow Parsnip stems can be dangerous if the stem is fresh or if you mistakenly use Douglas's Water-hemlock or Poison Hemlock, which are violently poisonous (as are several members of the celery family). Douglas's Water-hemlock and Poison Hemlock plants are more slender than Cow Parsnip and have smaller flower heads and finely divided leaves. Still, it is possible to confuse these species with Cow Parsnip, especially for inexperienced observers in the spring before the plant is fully developed. Cow Parsnip and its relatives contain phototoxic compounds, which make the skin sensitive to sunlight, so they must be handled very carefully. Even lightly touching your skin to the hairs on the leaves and then exposing it to sunlight can cause blistering and discoloration that may remain for weeks or even months. The effects can be especially severe for light-skinned people. (See Turner and Szczawinski 1991.)

Chocolate Tips
(Celery Family)

Lomatium dissectum
(Apiaceae)

Other Names: "Wild Celery", "Bitter-root".

Botanical Description
Chocolate Tips is a robust, multi-stemmed perennial, often growing a metre or more tall from a large, woody taproot. The leaves are large, especially the basal ones, and finely dissected into numerous small segments. The flowers are small and purple, clustered in large flat-topped umbrella-like heads. The fruits are elliptical and narrowly winged. *Lomatium dissectum* is also known as *Ferula dissecta* or *Leptotaenia dissecta.*

Habitat: dry rocky slopes from sea level to moderate elevations in the mountains.

Distribution in British Columbia: sporadic in the south, on both sides of the Cascade Mountains; common at Botanie Valley near Lytton and in parts of the Okanagan Valley.

Aboriginal Use
The Okanagan considered the tops and roots of Chocolate Tips poisonous, although they used to eat the new shoots in the spring, harvesting them before they emerged from the ground. Other Interior Salish peoples ate the young roots as well. The Okanagan, as well as the Spokane, Sahaptin and other Washington peoples, made a fish poison and insecticide from the roots. They pounded the roots and steeped them in water overnight to make a milky-coloured infusion. Then they poured the liquid into a creek, and the poisoned fish floated to the surface, where women and children could gather them. The poison lost its effectiveness about a kilometre downstream. Fish poisoned with Chocolate Tips were not harmful to eat as long as they were consumed soon afterward. The same solution poured over horses and cattle would rid them of lice and other insect pests. Rubbing the animals with the leaves and stems of Chocolate Tips achieved the same results.

"Indian Celery"   *Lomatium nudicaule*
(Celery Family)   (Apiaceae)

Other Names: "Indian Consumption Plant", "Wild Celery", Barestem Lomatium.

Botanical Description

"Indian Celery" is a herbaceous perennial whose solitary or clustered stems grow 20 to 60 cm tall from a stout taproot. The leaves are thick, bluish-green and compound, divided into 3 to 30 oval to lance-shaped leaflets, which may or may not be toothed at the tips. The flowers are light yellow, small and numerous, in loose umbrella-like clusters, with rays of varying lengths. The fruits are elliptical, winged and flat.

Habitat: dry open slopes, moist meadows and sparsely wooded areas, from sea level to moderate elevations in the mountains.

Distribution in British Columbia: in the south, both west and east of the Cascade Range; common on Vancouver Island and the Gulf Islands, as well as at Botanie Valley and many other interior locations.

Aboriginal Use

The spicy and aromatic seeds of "Indian Celery" were widely used as an incense, fumigant and house deodorant. The Saanich, Cowichan, Kwakwa̱ka'wakw, Squamish, Lower Stl'atl'imx and Secwepemc all used them for this purpose. The Saanich and Squamish placed them on a hot stove, allowing the smoke to permeate the air. The Saanich also used them as a flavouring for meat and fish, and burned them in the hut where salmon was drying, to show respect for the spirits of the salmon. The Lower Stl'atl'imx burned "Indian Celery" seeds in an open fire as an incense and a mosquito repellent. The Secwepemc placed the seeds under the mattress of a baby's basket as a scent, and also under the pillow of an older person to disinfect and deodorize the bed; it was said to act like baby powder.

"Indian Celery" seeds.

The Nlaka'pamux used the leaves of a related species, Large-fruited Desert-parsley ("Wild Carrot"), as a scent and as padding in child carriers to make the child sleep well. The Flathead of Montana used yet another *Lomatium* species, Narrow-leaved Desert-parsley, in perfume bags as a scent, and the Blackfoot of Alberta stuffed the seeds of this species into animal pelts during the tanning process.

Indian Hemp	*Apocynum cannabinum*
Spreading Dogbane	*Apocynum androsaemifolium*
(Dogbane Family)	**(Apocynaceae)**

Botanical Description

Indian Hemp is an erect, bushy herbaceous perennial that grows up to 1 metre tall, with smooth, often reddish stems. It has many opposite,

Indian Hemp.

finely pointed, elliptical to lance-shaped leaves, 5 to 11 cm long; they are yellowish green, turning golden yellow in the fall. A milky latex exudes from the stems when broken. The flowers, which bloom in summer, are small, bell-shaped, and whitish; they grow in clusters near the ends of the stems. The fruits, growing in pairs, are slender and 12 to 18 cm long, splitting open along one side when ripe to reveal numerous small, brown seeds, each with a long tuft of cottony hairs to aid in dispersal. Spreading Dogbane is similar to Indian Hemp, but with shorter stems, oval leaves that droop, and fragrant, showy pink flowers. The two species occasionally hybridize where their ranges overlap.

Habitat: both plants grow in open areas, along roadsides and in clearings, often forming dense patches; Indian Hemp grows best in damp hollows near rivers or sloughs, whereas Spreading Dogbane grows in open woods and gravely areas.

Spreading Dogbane.

Distribution in British Columbia: Indian Hemp is common in the valleys and lower slopes of the southern interior, while Spreading Dogbane grows from sea level to moderate elevations in the mountains in many locations throughout the province except on Haida Gwaii.

Aboriginal Use

Indian Hemp was without doubt the most important source of plant fibre for First Peoples of the southern interior. Spreading Dogbane was occasionally used when Indian Hemp was not available, but being shorter and bushier, it was a poor substitute. Okanagan people harvested Indian Hemp stems in September or October, just as the leaves were turning yellow. Damp areas were said to produce the tallest, thickest plants. The harvesters removed the branches and leaves, then flattened the stems by pulling them over a pole tied to a tree. They split open the stems from bottom to top with a knife or sharp stick and peeled off the outer "bark" (skin) by hand, breaking away the brittle inner tissues, then bundled the fibrous parts together and hung them by the tops to dry in the wind. When dry, the fibre can be easily separated from any remaining outer skin by pounding the flattened pieces with a stick or twisting them with the hands. The final step was to form the light-brownish-to-greyish fibres into twine by rolling them with dampened hands on the bare thigh or on a piece of buckskin draped

Bundles of Indian Hemp fibre (Stl'atl'imx): top, raw (RBCM 1217); bottom, processed (RBCM 4813).

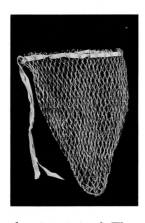

A Nlaka'pamux netted carrying bag made from Indian Hemp fibre, with a buckskin drawstring. (RBCM 2651)

over the leg while holding one end of the fibre to maintain the tension.

People joined lengths of fibre together by splitting the thick end of one piece and the thin end of another about one half the length of the stem and splicing them as an interlocking "V", then rolling them together until they intertwined. The splicing process could continue almost indefinitely; some people produced balls of twine as large as basketballs. An average plant yields about 75 cm of fibre, but nearly half of this is lost in splicing. A finer twine could be produced by splitting the stems in two along the entire length, breaking off the brittle inner part, and splicing and rolling the half-stem fibres together in the same manner as the whole pieces. Strong ropes could be made by twisting or plaiting two or more strands of twine together. A good, several-ply Indian Hemp rope is said to have the equivalent strength of a modern rope of a few hundred kilograms test weight. Even the thinnest of threads is difficult to break with the hands. When stored properly, Indian Hemp fibre will keep for many years without deteriorating. Its natural colour is a light tan, almost white.

The Okanagan and other interior peoples, including the Nlaka'-pamux, Stl'atl'imx, Secwepemc and Ktunaxa, used Indian Hemp twine for fishing lines and nets, because it keeps its strength under water and will not shrink. They also used it to make deer nets, slings, bowstrings, bridle ropes, nooses for game birds and hide stretchers, to bind implements, to sew moccasins, clothing, baskets, birch-bark canoes and Cattail mats, and to weave tumplines, garments, baby bedding and bags. They often wove Indian Hemp with other plant fibres, such as Tule stems and the bark of Silverberry, willow and sagebrush; in making garments, they sometimes spun it with deer hair. The Stl'atl'imx treated Indian Hemp fishing lines with a mixture of Lodgepole Pine pitch and Black Bear grease to prevent them from kinking. They sometimes coloured the lines with green leaves or alder bark dye to make them less conspicuous to fish. Loose Indian Hemp fibre was used in some areas as a tinder for starting fires.

Raw or spun, Indian Hemp fibre was a common trading product, not only among the aboriginal groups of the southern interior, but between them and the coastal peoples, as far west as Vancouver Island. In the Okanagan area, in the early days of contact with Europeans, a large bundle of prepared twine was worth as much as a horse. On the coast, the twine was a prized material for net-making, superior even to Stinging Nettle twine. The Carrier and other Athapaskan peoples to the north probably acquired a certain amount of Indian Hemp fibre through trade, but for the most part they had to content themselves with Spreading Dogbane fibre, which they prepared by a similar process. The Gitxsan used Dogbane fibre to make cord for fishnets and snares, and for the short crosswise element of woven pack straps. The Nuxalk also used the bushy Dogbane plants to fan the face on hot summer days.

Devil's Club
(Ginseng Family)

Oplopanax horridus
(Araliaceae)

Botanical Description
Devil's Club is a low, sprawling shrub, 1 to 3 metres high, covered with needle-like spines that can inflict painful wounds to those who touch them. It has greyish-brown bark and its wood is soft, with a distinctive sweetish odour. The leaves are large and maple-like, with seven to nine shallow, pointed lobes and toothed margins; in good sites they may grow to 30 cm or more across. The leaf stems and veins are also covered with sharp spines. The flowers are small and whitish, in compact heads arranged in pyramidal terminal clusters. The fruits are bright-red, spiny berries.

Habitat: moist, shady coniferous woods, often along stream banks and seepage faces in rich, black soil, from sea level to subalpine elevations.

Distribution in British Columbia: common along the coast from Vancouver Island to Alaska, and recurring in the interior wet belt and as far north as the Fort Nelson River.

Aboriginal Use

The Haida used Devil's Club stems to hook octopus from their dens. The Clallam of Washington and some Vancouver Island Nuu-chah-nulth groups peeled the sticks and cut them into small pieces for use as fish lures – the Manhousaht Nuu-chah-nulth and Ditidaht actually carved them to resemble small fish. Attached to fishing lines near the hook, they would spin to the surface underwater. The fish, attracted by the moving object, would unwittingly become ensnared on the hook. The Ditidaht sometimes used the wood to make fish-shaped lures and winglike propellers for cod lures. The Straits Salish, Squamish, Ditidaht, Haisla and other peoples used to mix Devil's Club charcoal with bear grease to make a black face paint for ceremonial occasions, such as winter dances. Some people even inserted it under the skin as a bluish-coloured tattoo. Erna Gunther (1945) reported that the Lummi were still using Devil's Club charcoal for face paint in the 1940s, but that they were mixing it with Vaseline instead of grease. This use continued into the 1990s in some areas. The Dena'ina of Alaska also used Devil's Club charcoal, mixing it with water to make a black dye.

Showy Milkweed.

Showy Milkweed
(Milkweed Family)

Asclepias speciosa
(Asclepiadaceae)

Other Name: Milkweed.

Botanical Description

Showy Milkweed is a herbaceous perennial, 40 to 120 cm tall, and usually bushy and branching. The leaves are opposite, smooth-edged, lance-shaped to oval, up to 20 cm long and about half as wide, with conspicuous transverse veins. The stems and leaves are covered with soft, woolly hairs that give the plant a greyish cast. The broken stems exude a white, sticky latex, giving the plant its common name. The flowers are light to dark pink, numerous and showy, in one or more umbrella-shaped clusters. The fruits are conspicuous, light green, broadly spindle-shaped capsules, 10 cm or more long, and covered with soft, recurved spines. They split open along one side when ripe to reveal numerous flat, brown seeds, each with a long tuft of white, silky hairs that aid in dispersal.

Habitat: dry roadsides to moist meadows and stream banks in sandy to loamy soil.

Distribution in British Columbia: east of the Cascade Mountains throughout the southern interior.

Aboriginal Use

The Nlaka'pamux and other Interior Salish groups sometimes used the stem fibre of Showy Milkweed as a substitute for Indian Hemp fibre to make twine for binding and tying. But they considered it inferior in quality and employed it only when Indian Hemp was not available. People harvested the stems in the fall, as they did those of Indian Hemp, and prepared them similarly. Some Okanagan people call Milkweed "Coyote's Indian Hemp" as a nickname, alluding to a mythical event in which Coyote transformed some Milkweed plants into Indian Hemp by urinating on them.

Pasture Wormwood	*Artemisia frigida*
Dragon Sagewort	*Artemisia dracunculus*
Western Mugwort	*Artemisia ludoviciana*
(Aster Family)	(Asteraceae)

Other Names: wormwood, mugwort, sagebrush (any *Artemisia* species); Prairie Sagewort (*A. frigida*); Wild Tarragon (*A. dracunculus*); Prairie Sage, Cudweed Sagewort (*A. ludoviciana*).

Botanical Description
Pasture Wormwood is a low, mat-forming aromatic perennial with a branching woody crown. The flowering stems are 10 to 40 cm high. The leaves are small and numerous, clustered at the base and spaced along the stems; they are finely dissected into many narrow segments. The minute, composite flowers are compacted in small, yellowish discs that grow in elongated, irregular terminal clusters. The stems, leaves and flower heads are covered with a dense mat of white, silky hairs, giving the entire plant a silvery appearance. Dragon Sagewort is also a perennial, with several tall stems ascending from a stout rhizome. It can be strongly aromatic or nearly odourless, smooth or slightly hairy, but never densely woolly. The leaves are long and slender, sometimes deeply cleft, the lower ones drooping with age. The flower heads are minute and green, in loose, compound terminal clusters. Western Mugwort is a tall, strongly aromatic perennial with silver-woolly leaves and stems. The leaves are lance-shaped and, in one variety (var. *latiloba*), sharply lobed. The flower heads are small, in compact but elongated terminal clusters.

Pasture Wormwood.

Habitat: dry open places in sandy or gravely soil; Western Mugwort grows along rivers and on gravel bars.

Distribution in British Columbia: all are found in dry locations east of the coastal mountains. The range of Pasture Wormwood extends north into the Peace River country, while the other two are confined to the southern half of the province.

Aboriginal Use

All three of these *Artemisia* species were valued for their aromatic fragrance, which acts as an effective insect repellent. The Okanagan, Secwepemc, Ktunaxa and the Blackfoot of Alberta burned branches of Pasture Wormwood and Dragon Sagewort as a smudge to drive away mosquitoes and other biting insects. They also placed the branches under pillows and mattresses to get rid of bedbugs, fleas and lice. The Okanagan, Secwepemc and Blackfoot employed Western Mugwort similarly. Other aromatic species of *Artemisia* were also used in areas where they occurred.

The Secwepemc covered the floors of their sweat-houses with a pleasant smelling mixture of Pasture Wormwood branches and Douglas-fir boughs. The Nlaka'pamux often used Pasture Wormwood branches for smoking hides, since they burn with a thick, heavy smoke. The Secwepemc used Dragon Sagewort branches as matting to sit on; the Okanagan used them to pad baby boards, cradles and diapers – they were said to keep the baby cool on a hot day, and heal diaper rash and rawness of the skin. The Okanagan also used the branches with the leaves still attached to make salmon spreaders for drying and storing salmon; the strong smell repelled flies and kept them from laying eggs in the flesh. The Flathead of Montana used the foliage of Western Mugwort, along with Douglas-fir boughs, in their sweat-houses as an incense, and rubbed hides with it before they were soaked to prevent them from going sour. The Blackfoot used Pasture Wormwood and Western Mugwort as a deodorant for saddles, pillows, hide bags, quivers and moccasins, and also as "toilet paper" and for cleaning paint applicators. They rubbed hides with an infusion of the stems and roots of Pasture Wormwood or the roots of Field Wormwood as a curing agent.

Dragon Sagewort.

Big Sagebrush
(Aster Family)

Artemisia tridentata
(Asteraceae)

Other Name: Common Sagebrush.

Botanical Description

Big Sagebrush is an erect branching shrub up to 2 metres tall, with grey, shredding bark and an aromatic odour. The leaves and young twigs are covered with a dense mat of fine, silvery hairs, giving the plant a soft pale-grey appearance. The leaves, 1 to 4 cm long, are mostly wedge-shaped with three rounded terminal teeth. They remain on the bushes during the winter. Flowering time is late summer and fall. The minute, greenish-grey flower heads grow in dense, branching terminal clusters.

Habitat: dry open plains and woods, extending to subalpine elevations in some locations.

Distribution in British Columbia: common throughout the dry southern interior; plentiful on overgrazed sites, since it is unpalatable to livestock.

Aboriginal Use

Like the other species of *Artemisia* mentioned, Big Sagebrush was valued by interior peoples for its aromatic scent. The Secwepemc and other groups placed the branches on a hot stove, as incense, to fumigate a house. They also boiled the foliage and used the solution for washing walls and floors, as a disinfectant and insect repellent. Some people used the leaves as a deodorizer when handling corpses.

All the Interior Salish groups pulled off the stringy, shredding bark of Big Sagebrush stems and (according to one source) roots in long strips and used it to weave mats, bags, baskets, quiver cases, saddle blankets, dresses, skirts, aprons, loin cloths, ponchos, capes, and even socks and shoes. But it was said that only poorer people, who did not have access to skins or other material, made clothing out of Big Sagebrush bark. The people who used the bark often interwove it with other fibres, such as willow bark, Red-cedar bark, Silverberry bark or Indian Hemp. The Okanagan stuffed pillows with Big Sagebrush bark, especially for a child's cradle. The Nlaka'pamux made quiver cases by binding the twigs together in a cylinder and sewing a piece of hide over the end. The Stl'atl'imx used the roots, hollowed out, to make temporary pipes.

Big Sagebrush wood was often used as a fuel, for cooking and smoking hides; it burns easily and is plentiful in the lowlands and river benches of the dry interior. The shredded bark made excellent tinder and was commonly used – as Red-cedar bark was on the coast – for making a "slow match" to carry on journeys. The Stl'atl'imx macerated the old dry bark, formed it into a tight ball and bound it with birch-bark, while the Okanagan twisted it into a rope up to a metre long. In either form, when ignited the bark would smoulder for a long time; it could be transported easily from place to place and used at any time to kindle a fire.

Nlaka'pamux child's sock of woven Big Sagebrush bark. This sock was worn inside a moccasin.
(RBCM 2749)

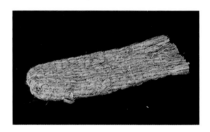

Balsamroot
(Aster Family)

Balsamorhiza sagittata
(Asteraceae)

Other Names: Spring Sunflower, "Wild Sunflower".

Botanical Description

Balsamroot is a perennial that grows 30 to 50 cm tall from a thick, deep-seated taproot. It has numerous large, long-stemmed leaves that grow in clusters; they are shaped like arrow heads and appear grey or silvery because of a thick covering of fine white hairs. The flower heads, usually several per plant, grow on individual stems; they are bright yellow and resemble small sunflowers with about 25 petal-like ray flowers

per head. The fruits, which shake loose easily from the old dried flower heads, resemble miniature sunflower seeds. Balsamroot blooms from April to July, depending on the elevation, at which time the plants create a striking display of colour on the hillsides and valleys of the southern interior.

Habitat: open dry hillsides and flats, from lowlands to moderate elevations in the mountains; particularly prevalent in overgrazed areas.

Distribution in British Columbia: widespread and abundant throughout the drier regions of the southern interior.

Aboriginal Use

In winter, the Okanagan stuffed the large, hairy leaves of Balsamroot in their moccasins to keep their feet warm. (They also used Pinegrass or deer hair for this purpose.) Young Okanagan boys training to acquire supernatural power would wrap the leaves around their feet, pinning them on with Bluebunch Wheatgrass stems, and walk on them to see how far they could get without tearing them. This exercise prepared them for walking silently and carefully in the woods. Some people said that gifted boys could even jog with the leaves on their feet.

Vanilla-leaf
(Barberry Family)

Achlys triphylla
(Berberidaceae)

Other Names: Sweet-after-death, May Leaves, Deer-foot.

Botanical Description

Vanilla-leaf is a herbaceous perennial, usually 15 cm or more tall, with a single long-stemmed light-green leaf ascending from ground level. The leaf has three leaflets – one terminal and two lateral – all more-or-less lobed or wavy-edged. The flowers are small and white, lacking petals or sepals. They grow in a compact, bottle-brush-like cluster at the end of a thin, wiry stem arising from the ground at the same point as the leaf. The fruits are small, rounded and greenish or reddish purple.

Habitat: moist, shaded woods, often growing in dense patches.

Distribution in British Columbia: along the southern coastal region, from the east slopes of the Cascade Mountains westward; common on southeastern Vancouver Island.

Aboriginal Use

The dried leaves of Vanilla-leaf have a faint vanilla-like odour, as implied by the common name. The Saanich, Nlaka'pamux and probably other groups on the southern coast, used the leaves as an insect repellent. The Saanich dried them and hung them in bunches in houses to keep flies and mosquitoes away. The Nlaka'pamux boiled the leaves and used the solution to bathe the skin of people infected with lice. They also washed bedding, furniture and floors with Vanilla-leaf water to eliminate bedbugs and other insect pests. This solution was also said to be effective against skin parasites on sheep.

Tall Oregon-grape
Dull Oregon-grape
(Barberry Family)

Mahonia aquifolium
Mahonia nervosa
(Berberidaceae)

Other Names: barberry, mahonia (both); Tall Mahonia (*M. aquifolium*).

Botanical Description
These species of Oregon-grape are low shrubs with leathery holly-like compound leaves, elongated clusters of bright yellow flowers, and long clusters of round, deep-blue berries that have a greyish waxy coating. The bark is light yellow-grey outside and bright yellow inside. As the name suggests, Tall Orgeon-grape is the taller of the two, sometimes exceeding 2 metres, and has 5 to 7 leaflets per leaf. Dull Oregon-grape usually has 9 to 15 leaflets per leaf. Some botanists include these species in the genus *Berberis*. A short subspecies of *Mahonia aquifolium* is sometimes called Creeping Oregon-grape (ssp. *repens*).

Tall Oregon-grape.

Habitat: Tall Orgeon-grape grows in dry open rocky areas, while Dull Oregon-grape prefers light to shaded coniferous forest.

Distribution in British Columbia: Tall Orgeon-grape occurs throughout the southern part of the province; Dull Oregon-grape is confined to the southern coastal forests west of the Cascade Mountains.

Aboriginal Use
The inner bark of the stems and roots of Oregon-grape (both species) contains a bright yellow pigment that can be extracted simply by boiling the bark in water. The Cowichan, Straits Salish and Nuu-chah-nulth of British Columbia, and the Chehalis, Skagit and Snohomish of Washington all used both species for dyeing basket materials, while the Nlaka'pamux and Okanagan used Tall Orgeon-grape, the only type available in their area.

To make the yellow dye, people boiled a handful of shredded sticks and roots in about a litre of water. Then they steeped the material they wanted to dye in the solution. They often coloured Bear-grass by this method when using it in basketry. As of the 1970s, the Upper Skagit have been using Oregon-grape to dye rags for making braided rugs.

The Okanagan dyed Porcupine quills with Oregon-grape. They also made a thick, concentrated paint from it by boiling the dye solution until the water evaporated, leaving a yellow powdery substance, then mixing the powder with ochre paint or Cottonwood bud resin. They often intensified the colour of the paint by boiling Wolf Lichen with the Oregon-grape.

Some Vancouver Island Salish groups used an extract of Oregon-grape roots as a detergent lotion for washing the hands.

Red Alder	*Alnus rubra*
Mountain Alder	*Alnus tenuifolia*
Green Alder	*Alnus crispa* ssp. *sinuata*
(Birch Family)	**(Betulaceae)**

Other Names: Oregon Alder (*A. rubra*); Sitka Alder, Slide Alder (*A. crispa*).

Botanical Description
Red Alder is a fast-growing straight-trunked deciduous tree that can reach 25 metres tall with a trunk diameter of 80 cm. The bark is smooth and greenish when young, becoming coarse and grey or whitish with age. The inner bark and wood turn deep red or orange when exposed to air. The leaves are bright green on top and lighter underneath, oval and pointed, with coarse teeth. The male flowers hang down in long catkins, which release clouds of pollen when they mature in early spring. The female flowers are short egg-

Red Alder.

shaped cones that grow in clusters; they are green and resinous when immature, brown and woody when ripe. The fruits or nutlets are small and flat with short lateral wings.

Mountain Alder and Green Alder are similar to Red Alder, but generally lower, more bushy and shrublike, with more fincly-toothed leaves, often doubly serrated. Mountain Alder has blunt winter buds and its nutlets have no wings; Green Alder has sharply pointed winter buds and its nutlets have broad wings. *Alnus tenuifolia* is also known as *A. incana*, and *A. crispa* ssp. *sinuata* as *A. sinuata* or *A. sitchensis.*

Habitat: cool, moist woods, along stream banks and swamp edges; Red Alder is common in recently cleared areas; Green Alder and Mountain Alder often form dense patches on slide areas and avalanche runs in the mountains.

Distribution in British Columbia: Red Alder is common in forests west of the Cascade Mountains along the entire coast, and also occurs sporadically in the southern interior; Mountain Alder is found throughout the interior east of the Coast Mountains, except in the far northeastern corner of the province; Green Alder grows throughout the province.

Aboriginal Use

British Columbia First Peoples used alders extensively for dyeing and carving, and for fuel. First Peoples across the continent know of alders as a source of orange-red dye. Virtually every aboriginal group in the province that had access to Red Alder used it in this capacity. Red Alder was used along the coast and in some areas of the interior where it could be found; otherwise Mountain Alder or, occasionally, Green Alder was used. By varying the preparation techniques, people could produce colours ranging from almost black to dark brown to russet to bright orange-red from the wood and bark of alders. They used alder dyes to colour basket materials, cedar bark, ropes, fishnets and lines, wooden articles (canoes, canoe bailers, masks, rattles, tool handles and totem poles), Mountain Goat wool, feathers, Porcupine quills, human hair and buckskin. They even used the bark as a pigment for tattooing.

The simplest method of preparing alder dye was to boil the bark or wood (or both) in a small quantity of water, and then to steep the material to be coloured in the solution. This procedure usually yielded a reddish brown colour, suitable for fishnets and basket materials. The colour helped make nets and lines invisible to fish. The Haida, Nuu-chah-nulth and Kwakwaka'wakw produced a brighter red for cedar bark by chewing the alder bark, spitting the saliva into a container and bringing it to a boil by adding red-hot rocks.

The Kwakwaka'wakw, Tlingit and Nuxalk often used urine as a mordant to obtain a bright red dye. The Tlingit carved vessels out of Red Alder trunks and filled them with children's urine, allowing it to stand for a time until it absorbed the red colouring from the alder wood, then dipped the material to be dyed into the solution. The Nuxalk used the following procedure: In summer they used scrapings from the inner bark, in winter large pieces of the bark with the wood attached. They poured water into a vessel, then added urine and alder bark. Using red-hot rocks, they heated this mixture gradually to the boiling point, stirring it occasionally. When it boiled, they took out some of the bark and added more, and continued this process until the solution was a deep red. They allowed it to stand for a few minutes, then put in the cedar bark or other material, gently working it until it was saturated. Finally, they removed the material and hung it up to dry. Several pieces of cedar bark could be dyed in the same solution provided more bark and hot stones were added to maintain the proper strength and temperature. T.F. McIlwraith recorded this procedure in his book, *The Bella Coola Indians* (1948), stating that the Nuxalk (Bella Coola) still dyed cedar bark this way in the 1940s, even using hot rocks to heat the solution, because they believed that heating it on a stove would produce inferior results.

The Saanich mixed Red Alder bark with cedar bark and Indian Paint Fungus, burned them to a powder, and inserted this under the skin with a needle to make tattoos. They also put Red Alder bark in steaming pits to colour their Blue Camas bulbs pink. The Haida used the charcoal from the wood for tattooing and put about half a cup of the bark in their wash water as a bleach substitute. Several coastal groups have traditional narratives containing episodes in which the hero feigns bleeding at the mouth by chewing pieces of Red Alder bark and letting the saliva ooze from his lips – in this way, he fools his enemies into believing he is dead.

The interior peoples commonly used alder bark to treat animal hides. The Stl'atl'imx rubbed skins on peeled alder trees to darken their colour from yellowish to reddish-brown. The Okanagan simply rubbed skins with the fresh bark. Okanagan people made a concentrated red paint by boiling the wood in water for a long time, until the liquid had nearly evaporated, then adding a few drops of fish oil, stirring constantly, and finally removing it from the heat and powdering it on a piece of alder bark. This paint could be used on wood, hides or body. The Secwepemc and Nlaka'pamux steeped buckskins overnight in a cooled solution of alder bark, colouring and tanning them at the same time. The most commonly used pigment of the Secwepemc was the reddish-brown dye made from alder – they used it to colour gambling sticks, Porcupine quills, hair, feathers, straw, dressed skins and buckskin clothes. They simply soaked these items in an alder-bark solution, took them out and wrung them dry, then resoaked them until they obtained the desired shade. The Secwepemc sometimes mixed the bark with Saskatoon berries to make a dark purple dye for hides. They also made a black dye by boiling the bark with roasted iron pyrites. Secwepemc elder Mary Thomas described how Bitter Cherry bark was dyed black by peeling away the bark of a large alder root, inserting the cherry bark in the cut, closing it up and leaving it for several weeks. The Flathead of Montana used alder dye to tint moccasins yellow, feathers reddish brown and human hair a flaming red.

The importance of alder as a dye is well known, but the value of the wood is often overlooked. Alder wood makes excellent fuel, and is considered one of the best for smoking salmon and cooking deer meat, because it has a low pitch content and does not impart any unpleasant flavour to the food. Seasoned or partially rotten wood is said to be better for smoking than green, fresh wood. The Carrier put alder-bark chips in with the hot rocks at the bottom of steaming pits; they would burn for many hours with a slow, steady heat. The Okanagan cleaned their teeth with the burnt ashes of alder and birch.

The absence of pitchy flavour in the wood and its smooth, even-grained texture made it ideal for carving spoons, vessels and serving platters. Some of the finest examples of Northwest Coast bowls and feast dishes are made of Red Alder. It was also used to carve rattles, masks, adze handles, headdresses, frontlets for headdresses, arrow points, pendants, labrets, canoe bailers and paddles. Also, the Haida carved spoons, small dishes, masks and rattles from the smaller Green Alder, and the Ktunaxa made pipe stems from Mountain Alder twigs. The Gitxsan sometimes used Mountain Alder wood to make spoons, and used it for fuel. The Dena'ina of Alaska used Mountain Alder and Green Alder wood to construct shelters, fish traps and weirs. They also used the wood as fuel for smoking meat and fish, the bark as a dye for skins and snowshoes, the roots for twine, and the branches for steam-bath switches.

The Nlaka'pamux used the fragrant stems of Mountain Alder as a scent or perfume, and they sometimes used the young twigs for basket imbrication. The Okanagan made string from the bark of young alders, and occasionally substituted Mountain Alder roots for Red-cedar roots in making coiled baskets. They peeled, split and soaked the roots in water to make them pliable for weaving. The Carrier wove fishnets of alder bark, then dyed them black by boiling them in their own juice. People sometimes used alder branches as matting for cleaning salmon or for surrounding the food in steaming pits.

Paper Birch *Betula papyrifera*
(Birch Family) (Betulaceae)

Other Names: White Birch, Western Paper Birch, Canoe Birch.

Botanical Description
A small to medium deciduous tree, Paper Birch can grow up to 20 metres tall. The bark, when mature, is reddish-brown to chalky white, usually peeling readily in horizontal strips and separating into thin layers; the young twigs are usually hairy and often glandular. The leaf blades, 4 to 7 cm long, are oval to nearly round or slightly heart-shaped, sharply pointed, and coarsely or finely toothed. The leaf stalks generally exceed 15 mm in length. The flowers grow in separate male and female catkins, the former long and clustered, the latter shorter and usually single. The catkin scales are shed with the fruits, which are small with lateral wings. A highly variable species, Paper Birch intergrades freely in southern British Columbia with the closely related Water Birch (also Mountain Birch or Black Birch). The leaves of the Water Birch are not as sharply pointed and have shorter stalks; its bark is dark coppery-red to purplish-brown and does not separate easily into layers.

Habitat: moist open woods along streams and lake edges from valley bottoms to moderate elevations in the mountains.

Distribution in British Columbia: widespread throughout the interior, and also common in some areas of the coastal mainland; rare on Vancouver Island and not found on Haida Gwaii. At least two varieties are distinguished in the province.

Aboriginal Use

The bark of Paper Birch, which can be peeled off the tree in large, flexible sheets, was as important to the First Peoples of the interior as the bark of Western Red-cedar was to coastal peoples. It could be stripped off at any time of the year, but was said to peel most easily in late spring and early summer when the sap was running. Bark with short lenticels (horizontal lines) was better than that with long lenticels, because it would not split and crack as much when it was being worked on. People used only the bark of the Paper Birch; the bark of the closely related Water Birch was not of suitable quality and, being thinner, was more difficult to harvest.

To harvest the bark of the Paper Birch, people made two horizontal cuts around the tree, one high and one near the ground, and then a single vertical cut between them. They peeled off the sheet by lifting the edges along the cuts and pulling horizontally. When properly done, the harvesting did not kill the tree because only the outer bark was removed, not the innermost layer next to the living cambium tissue. Sometimes, however, people would cut down a tree to collect bark from the upper trunk. Paper Birch trees showing evidence of past harvesting (known as Culturally Modified Trees) are common in parts of British Columbia's interior.

Baskets and canoes were the items interior peoples most commonly made from birch bark. Some coastal peoples, including the Upper Sto:lo and Nuxalk, also made these on occasion, having learned the craft from their interior neighbours. Certain peoples, such as the Secwepemc, Gitxsan and Wet'suwet'en, are famous for their skill in working with birch bark. Their baskets were widely traded among the peoples of the interior and, today, are commonly sold in gift shops, such as the shop at the Secwepemc Museum in Kamloops.

Women constructed baskets by making four diagonal cuts, two from each edge, towards the middle of a rectangular sheet of bark. They folded the sheet into a boxlike shape, with the cuts directed towards the bottom corners and the edges coming together to form side seams. In accordance with the natural tendency of the bark to curl outward when peeled off the tree, the whitish outer surface of the bark formed the inside of the basket and the reddish-brown inner surface formed the

Secwepemc elder Mary Thomas holding birch-bark baskets she has made. She sewed the baskets with split cedar root and made the rims with Saskatoon Berry wood.

basket's exterior. The women sewed the side seams, usually with split-cedar roots, spruce roots or willow bark, and then bound or stitched to the top a circular hoop of the same material or of Saskatoon Berry, willow, cedar, Red-osier Dogwood or some other flexible wood. Finally they caulked the seams with pitch, and sometimes etched designs – some of them very intricate – on the outer surface. Some basket makers used strips of Bitter Cherry or Pin Cherry bark to make decorative patterns around the rims of the baskets. The women made birch-bark containers in a variety of sizes and used them in berry picking, for storing food, for boiling food with hot rocks and even for packing water. When cooking with hot rocks, they laid green sticks of Saskatoon Berry or some other shrub in the bottom of the basket to prevent the rocks from burning through the birch bark.

The men of interior groups could make a canoe from a single piece of birch bark: they folded and sewed the bark onto a frame of willow or cedar withes, then sealed the seams and cracks with pitch. Some of the canoes were 5 metres or more in length. Birch-bark canoes were strong yet buoyant, and with proper handling were capable of tremendous speeds. Some of the canoes made by Athapaskan peoples such as the Carrier were so skilfully constructed that they could be dismantled and folded for portaging.

People also used birch-bark strips to wrap food for storage, to line underground caches, to line graves and cover corpses, to splint broken limbs and bind implements, and as roofing and siding for temporary shelters. The Stl'atl'imx placed birch-bark funnels around the poles of raised food caches to protect them from climbing rodents. The Tahltan made snow goggles from the bark, the Dunne-za made Moose calls, and the Carrier made toboggans. The Stl'atl'imx, Nlaka'pamux,

Secwepemc and other groups made birch-bark infant carriers, cradles and urine conduits; birch-bark cradles are still treasured baby gifts today. The Dena'ina of Alaska made birch-bark hats, and also used the bark to make a dye for skins.

Birch wood – uniform in texture, strong and close-grained, but not very durable – was employed in a variety of ways. The Stl'atl'imx carved dishes, cups, spoons and digging-stick handles from it. The Carrier used it to make mauls, digging sticks and snowshoe frames, and the Tahltan for snowshoe frames and ground sticks, bows and gambling sticks. Birch-wood snowshoes were said to be excellent for dry snow, but absorbed moisture and became too heavy in wet snow. The Dunne-za sometimes made birch-wood arrows. The Dena'ina used Paper Birch to make snowshoes, sleds, boat frames, bows, arrows, spoons, dishes, tool handles, mauls, drums frames, fish traps, bowls and brooms; they also used the wood as a fuel and for construction. The Nisga'a, Gitxsan and Wet'suwet'en peoples still make birch-wood spoons and masks; they used to twist ropes from the roots for lashing fishing weirs. The Haida imported birch wood from the Nass to make seaweed chopping blocks, used in the traditional preparation of Red Laver. Some Ktunaxa people have recently used birch wood for smoking bacon. The Okanagan, Secwepemc and other interior groups used both Paper Birch and Water Birch as a general fuel. They considered the bark, shredded and paper thin, an excellent fire starter; the Gitxsan used it for torches

The Secwepemc steeped birch leaves in water to make a shampoo, and mixed birch leaves, children's urine and alkali clay from the edges of certain lakes to make soap for washing the skin.

The Tsilhqot'in and other Athapaskan peoples have often used the

flexible twigs of Water Birch to make baby-carrying frames; they cover the frames with printed cloth. The Dena'ina and others used the leafy branches of Swamp Birch for padding on the back when carrying meat and as an underlay for cutting meat and fish.

Fire starters: shredded birch bark and Cinder Conk (for making a "slow match"; see page 55).

Hazelnut
(Birch Family)

Corylus cornuta
(Betulaceae)

Other Names: Wild Filbert, Cobnut.

Botanical Description
Hazelnut is a bushy shrub, usually 2 to 5 metres tall, spreading and pro-
fusely branching. The young twigs are woolly. The leaves are broadly
oval, pointed and sharply toothed. The male flowers are borne in long,
yellowish catkins, ripening in early spring; the female flowers are small
and bright red, growing at the tips of the twigs. The nuts grow alone or
in clusters of two or three towards
the ends of twigs. They are en-
cased in long, tubular husks that
are light green and covered with
stiff, prickly hairs. When ripe, the
nuts resemble commercial filberts.

Habitat: from shaded forests on
the coast to open moist or rocky
areas in the interior.

Distribution in British Columbia: widespread throughout the southern part of the province from Vancouver Island to the Kootenays, and
extending well into the northern interior.

Aboriginal Use
The Straits Salish and Lower Stl'atl'imx used the young, straight sucker
shoots of Hazelnut to make arrows. The Upper Sto:lo of the Fraser
River valley, the Lower Stl'atl'imx, and the Chehalis and Skokomish of
Washington, peeled Hazelnut shoots, twisted them until soft and pliable, and used them alone or as a three-ply rope for tying and lashing.
In the interior, the Secwepemc and Gitxsan used the fresh branches as
matting for cleaning salmon on and for sitting and sleeping on. The
Secwepemc used the suckers as edging for birch-bark baskets and cradles, and made spoons from the wood because it has no strong flavour.
A hazelnut branch with a secondary twig still attached was sometimes
used by the Secwepemc as a fish hook. The Gitxsan made hockey sticks
from bent Hazelnut roots. The Okanagan sometimes used Hazelnut
saplings to make fish traps, tying them together with Indian Hemp

string. A blue dye, used to colour basket materials, was obtained from the roots by the Nlaka'pamux, and from the inner bark by the Sanpoil-Nespelem Okanagan of Washington, by steeping these parts in water.

Lemonweed   *Lithospermum ruderale*
(Borage Family)   (Boraginaceae)

Other Names: stoneseed (all *Lithospermum* species); Gromwell, Yellow Puccoon, Indian Paint (*L. ruderale*).

Botanical Description
Lemonweed is a herbaceous perennial with leafy stems up to 45 cm tall, ascending from a woody taproot. The leaves are linear or lance-shaped, 2 to 6 cm long, numerous, and crowded along the stems. Both stems and leaves are covered with stiff, sharp hairs. The flowers are light yel-

low, tubular and showy, sprouting from the axils of the uppermost leaves, or, later in the season, on the lower portion of the stems. The fruits are smooth nutlets that are hard, pointed and shiny, resembling white teeth. A similar species, Long-flowered Stoneseed, is somewhat shorter, with larger, bright yellow flowers and pitted nutlets.

Habitat: dry open plains, foothills in rocky or gravely soil, and upland meadows such as at Botanie Valley.

Distribution in British Columbia: the dry southern interior.

Aboriginal Use
The Okanagan used Lemonweed roots as a red dye. The Secwepemc reportedly used a red fluid from the stems to make a deadly poison for arrow tips, said to kill a person at a touch. This fluid, said to be very

rare and difficult to obtain, was steeped in hot water to produce the poison. Okanagan fishermen drew their lines through a handful of this plant to give them good luck in fishing. It probably had the effect of masking any human odour on the lines. Children used the shiny white nutlets as beads.

The Nlaka'pamux, Stl'atl'imx, Secwepemc and the Blackfoot of Alberta made a red paint from the red-tipped roots of Long-flowered Stoneseed; they used the paint to draw designs and pictures on dressed skins, gambling sticks, bows and faces. They dipped the roots in hot grease and painted on the colour directly. When first applied the paint is blood red, but it fades with age to a dull purple or violet. The Stl'atl'imx and probably other groups fixed the colour by rubbing it with the heated stems of prickly-pear. The Nlaka'pamux name for Lemonweed means "bloody". Puccoon, one of the common names for *Lithospermum*, was derived from an Algonkian term for plants used for staining and dyeing.

The Blackfoot dried the tops and seeds of Lemonweed as an incense.

Brittle Prickly-pear　　　　　　*Opuntia fragilis*
Plains Prickly-pear　　　　　*Opuntia polyacantha*
(Cactus Family)　　　　　　　**(Cactaceae)**

Botanical Description
Prickly-pears are low-growing perennials, often spreading into mats several metres broad. They have jointed, succulent stems covered with clusters of a few rigid spines, up to 5 cm long, arising from cushions of numerous short bristles. The flowers are 5 to 7 cm long, yellow, sometimes turning pinkish with age, with many petals and stamens – they are very showy. The fruits are small, berrylike, reddish and spiny. The stem segments of Brittle Prickly-pear are round in cross-section and usually 2 to 5 cm long, while those of Plains Prickly-pear are flat and mostly 5 to 15 cm long.

Habitat: dry hillsides and open plains.

Distribution in British Columbia: common in the dry valleys and hillsides of the southern interior; the range of Brittle Prickly-pear extends north into the Peace River region and it also occurs on the rocky points of southern Vancouver Island and the Gulf Islands.

Aboriginal Use

The Okanagan placed rings of cactus around the supporting poles of raised caches to keep mice and other animals from climbing up. They used the sharp spines to pierce ears and, when bones for fish-hooks were unavailable, they joined two cactus spines together in a V-shape or four together in a cross to make a temporary hook. They tied the spines with Indian Hemp string and sealed the joint with pitch. The Secwepemc also made fish hooks in this manner. The Stl'atl'imx and the Blackfoot of Alberta fixed painted designs on wood and buckskin with the mucilaginous juice from the inside of cactus stems, and the Nlaka'pamux used the juice to make face paints stay on longer. They heated freshly cut stems and rubbed them over surfaces painted with

Brittle Prickly-pear.

vegetable pigments (such as Lemonweed root) or with mineral ochre colours. The Blackfoot also employed the sticky juice to clear muddy water. They placed a cut cactus stem in a container of the water and agitated the mixture, allowing the silt particles to become entrapped in the mucilage of the cut stem. Nlaka'pamux women wore necklaces of the flat, disc-shaped cactus seeds.

Orange Honeysuckle
(Honeysuckle Family)

Lonicera ciliosa
(Caprifoliaceae)

Botanical Description

A shrubby vine, Orange Honeysuckle twines and climbs on other shrubs and trees, sometimes to a height of 6 metres or more. The bark is light greyish-brown and stringy. The leaves, in opposite pairs, are elliptical, pointed or rounded, and smooth-edged; the terminal pair of leaves is fused into a single concave disc or cup, forming the base of the flower cluster. The undersides of the leaves and young stems are covered with a whitish waxy coating, giving them a light blue-green colour. The tubular flowers, in tight clusters, are bright orange and showy. The fruits are soft, fleshy, red berries; they are not edible. Flowering time is in the summer; the berries ripen in August and September.

Habitat: roadside thickets and open woods from sea level to moderate elevations in the mountains.

Distribution in British Columbia: common in the southern parts of the province, especially west of the Cascade Mountains.

Aboriginal Use

The Nlaka'pamux and possibly other Interior Salish peoples used the stem fibre of Orange Honeysuckle, usually interwoven with other types of fibre, to make mats, bags, capes, aprons and blankets. They employed the woody vines for reinforcing suspension bridges over the Fraser, Thompson and other rivers. The Flathead of Montana boiled the stems to make a shampoo, which they said made hair grow longer.

Black Twinberry
(Honeysuckle Family)

Lonicera involucrata
(Caprifoliaceae)

Other Names: Twinflower Honeysuckle, Bearberry Honeysuckle, Fly Honeysuckle, "Bearberry".

Botanical Description
A bushy shrub, Black Twinberry grows up to 3 metres or more tall. It has light brown, shredding bark, and paired, pointed, elliptical leaves. The flowers are yellow and tubular; they grow as twins within pairs of thin, leafy bracts, which turn bright red or purplish as the fruits mature.

The fleshy twin berries are black, shiny and beadlike, and are not good to eat. Flowers may be found on the bushes from April through August, with berries ripening from July to September.

Habitat: moist thickets and open, swampy areas from sea level to subalpine elevations.

Distribution in British Columbia: widespread throughout the province, especially on the coast and in the interior wet belt; sporadic but locally abundant.

Aboriginal Use
Several groups used the purple juice from Black Twinberry berries as a dye. The Secwepemc used it to dye roots for basketry, and the Quileute of Washington used it to paint dolls' faces. The Kwakwaka'wakw mashed the berries with Salal berries to intensify the colour. The Haida rubbed the berries into their hair as a tonic to prevent it from turning grey. They sometimes made gambling sticks from the wood.

Red Elderberry
Blue Elderberry
(Honeysuckle Family)

Sambucus racemosa
Sambucus cerulea
(Caprifoliaceae)

Other Names: Red Elder, Blue Elder.

Botanical Description

Elderberries are large, bushy shrubs, some-times treelike, 2 to 5 metres tall. They have grey-brown bark and brittle pithy twigs. The leaves are compound, each bearing five to nine pointed, toothed, lance-shaped to oval leaflets. The flowers are small and creamy-white. They grow in large clusters that are pyramidal in Red Elderberry and flat-topped in Blue Elderberry. Red Elderberries are bright red (or black in one form) and Blue Elderberries are deep blue with a waxy whitish bloom. *Sambucus racemosa* is also known as S. *pubens*, and S. *cerulea* is also called S. *glauca*.

Habitat: Red Elderberry occurs in open swampy areas, moist clearings and shaded woods, while Blue Elderberry is found on valley bottoms and open dry slopes; both grow from sea level to moderate elevations in the mountains.

Distribution in British Columbia: Red Elderberry is widespread along the coast, extending into the interior along some of the major river valleys, and recurring in the interior wet belt along the Columbia River, where its dark-colour form is common; Blue Elderberry is abun-dant in the southern interior, in the Okanagan and Columbia River val-leys, and sporadic on southeastern Vancouver Island and the Gulf Islands.

Aboriginal Use

Elderberry stems are easy to hollow out, and many aboriginal peoples used them as whistles, drinking straws, blowguns and pipe stems. But they are poisonous, as are the roots and foliage, so using them for such items is not recommended, especially when they are fresh. The Straits

Salish made pipe stems and the Nuxalk pipe bowls from Red Elderberry, and the children of these and other groups, such as the Kwakwaka'wakw, made blowguns similar to the modern pea shooter. They used small pieces of kelp or other vegetation as ammunition. The Stl'atl'imx made blowguns from the stems of both species; the Okanagan used only Blue Elderberry stems, from which they also made drinking straws for use by girls at puberty. The Sanpoil-Nespelem Okanagan of Washington and the Flathead of Montana made small flutes, the former from Blue Elderberry, the latter from both species. The Quinault of Washington made Elk whistles from Blue Elderberry stems and the Stl'atl'imx sometimes made urine conduits for baby carriers from Red Elderberry, and probably from Blue Elderberry as well, when rolled birch bark was not available.

The Haida used Red Elderberry pith to fasten flint tips onto arrow shafts, while the Kwakwaka'wakw, Haisla and others fixed segments of the stem onto arrows used to stun birds and squirrels. The Kwakwaka'-wakw also used pieces of the stem as bases for feather shuttlecocks that they employed in a game. The Mainland Comox made 15-cm long Lingcod lures from pieces of hollowed-out Red Elderberry stem, with several seagull feathers inserted at one end. The Sanpoil-Nespelem Okanagan sometimes used the straighter branches of the Blue Elderberry to make temporary arrows for hunting small game.

Common Snowberry
(Honeysuckle Family)

Symphoricarpos albus
(Caprifoliaceae)

Other Name: Waxberry.

Botanical Description
Common Snowberry is an erect, bushy shrub, 1 to 2 metres tall, with greyish bark. The leaves are opposite, elliptical or oval, mostly 1.5 to 5 cm long, often irregularly lobed, especially those on new twigs and suckers. The flowers are small, urn-shaped, whitish or pink, in dense, few-flowered clusters. The fruits are berrylike, white, globular and fleshy, often remaining on the bushes over winter.

Habitat: thickets, woods, and open areas from sea level to moderate elevations in the mountains.

Distribution in British Columbia: widespread throughout the province.

Aboriginal Use
The Okanagan and the Secwepemc, the Blackfoot of Alberta, and probably other groups used Common Snowberry branches to make brooms. They bound the bushy twigs together in a tight bundle, and sometimes tied it onto a long stick for a handle. The Secwepemc and Gitxsan hollowed out the twigs and used them as pipe stems, for pipe bowls of flint or soapstone. The Haida used Common Snowberry sticks, peeled, trimmed and sharpened, to skewer clams, cockles and mussels on for drying. The Blackfoot used the slender branches to make arrow shafts. But the Ktunaxa considered them a bad material for arrows – in an episode in one of their traditional narratives, Coyote shows his ignorance by using Common Snowberry wood to make an arrow. The Blackfoot made a fire from green Common Snowberry twigs and used the smoke to blacken the surface of newly made pipes before they were greased and polished.

Flowering Dogwood
(Dogwood Family)

Cornus nuttallii
(Cornaceae)

Other Names: Pacific Dogwood, Western Flowering Dogwood.

Botanical Description
Flowering Dogwood is a handsome tree that grows up to 20 metres tall; or it can also be shrublike. The bark is smooth and greyish, and the branches are whorled. The leaf blades are oval to elliptical, pointed and smooth-edged, with prominent veins. The leaves turn a bright peach colour in the fall. The flowers are numerous, small and greenish, growing in compact, but-

tonlike heads, each surrounded by four to six large, showy, white bracts. The flower heads begin to form in the fall, maturing in the spring as the leaves expand. Some trees produce a second blooming of flowers in the fall. The fruits are bright red-orange, elongated and berrylike, in compact clusters. Flowering Dogwood is the provincial flower of British Columbia.

Habitat: moist, open to shaded woods and slopes, at low elevations.

Distribution in British Columbia: common on southern Vancouver Island and the adjacent mainland, eastward in the Fraser River canyon towards Lytton.

Aboriginal Use
The wood of Flowering Dogwood is hard and tough, but not durable. It was the main bow-making material of the Lower Stl'atl'imx in the Pemberton Valley – far more important in that area than Western Yew, according to modern consultants. It was also used by the Lower Stl'atl'imx for making arrows and combs. The Nlaka'pamux made bows and implement handles from it, and the Cowichan on Vancouver Island used it for making bows, arrows and, recently, knitting needles. In Washington, the Skagit, Clallam and Green River groups used it to make gambling discs, the Skagit to make harpoon shafts, and the Snohomish to make sticks for pounding Bracken Fern rhizomes. The

Straits Salish used the bark as a tanning agent and the Nlaka'pamux used it to make a deep brown dye. They also mixed it with Grand Fir bark to make a black dye for colouring Bitter Cherry bark, which was used in basket imbrication (see page 185).

Red-osier Dogwood (Dogwood Family)

Cornus sericea (Cornaceae)

Other Names: "Red Willow", Western Dogwood, Creek Dogwood.

Botanical Description
Red-osier Dogwood is a many-stemmed shrub, 2 to 8 metres tall, with smooth, greenish to bright red or reddish-purple bark. The leaves are oval to elliptical, smooth-edged and pointed, with prominent, evenly spaced lateral veins; they turn bright red in the fall. The small, white flowers grow in dense flat-topped clusters. The fruits are round and white, often with a bluish or greenish tinge; they are fleshy and have a hard stone in the centre. There are two varieties of *Cornus sericea*: var. *stolonifera* has berries with smooth stones and var. *occidentalis* has berries with grooved stones. The former is more common east of the Coast Mountains and the latter west of this range, but they tend to intergrade. *C. sericea* is also known as *C. stolonifera*.

Habitat: moist soil, in marshes and swamps, and along streams and lake edges.

Distribution in British Columbia: widespread throughout the province from sea level to near the timberline.

Aboriginal Use
The First Peoples in this region used the thin, flexible branches of Red-osier Dogwood in a variety of capacities. The Secwepemc and others used them as salmon stretchers and skewers for barbecuing salmon; the branches were said to impart a salty flavour to the flesh. The Secwepemc also used them to make the rims of birch-bark and cedar-root baskets, to line cooking pits, to make a grid to hold the food being cooked in a steaming basket, for the frames of sweat-houses, and to make fish

Red-osier Dogwood.

traps and weirs. They considered the wood a good fuel for smoking and drying meat and fish, because it doesn't blacken the meat; they also used it with Saskatoon Berry wood as fuel for drying huckleberries. The Okanagan used the branches to construct fish traps and weirs, and made spatulas, cooking grids and skewers from them as well. The Ktunaxa and Oweekeno used them for pelt stretchers, and the Flathead of Montana to construct sweat-houses. The Nuxalk made barbecue racks from them and the Haida used them to make drying racks and other types of frames. The Gitxsan used the wood for children's bows and the twisted branches for tying things; they used the leafy branches as matting for cleaning salmon. The Blackfoot of Alberta made pipe stems from the sticks. The Dena'ina of Alaska used the twigs for birch-bark basket rims.

The Okanagan and the Secwepemc used Red-osier Dogwood bark fibre as cordage for tying and lashing. They stripped off the bark and spliced the pieces together, or sometimes twisted the entire branch, because the wood added strength to the rope. They used the rope to bind underwater implements, to tie fish traps, barbecue sticks and smokehouse frame poles, and as latticework for fish weirs. The Okanagan mixed powdered Red-osier Dogwood bark with the resin of Cottonwood buds to make a red paint.

Silverberry
(Oleaster Family)

Elaeagnus commutata
(Elaeagnaceae)

Other Names: Silver Buffalo Berry, "Silver Willow", "Pink-barked Willow", Wolfwillow.

Botanical Description

Silverberry is an erect, bushy shrub that grows 1 to 4 metres tall. It has smooth grey-ish-brown to pinkish bark and silvery-scurfy leaves that are lance-shaped to elliptical, 2 to 7 cm long, pointed and smooth-edged. The flowers, which bloom in early summer, are small, yellow and tubular, growing in small clusters in the leaf axils; they have a distinc-tive, almost overpowering odour that some people enjoy and others consider disagree-able. The fruits are spherical, silvery and pithy with a hard, striped central stone.

Habitat: open gravely slopes and bench-lands, and along gullies and watercourses.

Distribution in British Columbia: most abundant along water-courses in the dry southern interior, but extends northward to Alaska and the Yukon.

Aboriginal Use

Silverberry bark is tough and fibrous and was an important weaving and rope-making material for the Interior Salish and the Ktunaxa. People stripped off the bark with a knife, usually in spring. Discarding the outer part, they soaked the stringy inner layers in water and spun them into twine on the bare thigh. Silverberry twine could be woven into bags, baskets, nets, mats, blankets and clothing, or plaited into rope for binding, tying or use as fishing line. The Nlaka'pamux and Okanagan were especially proficient in the preparation and utilization of this fibre. They braided strands of it and intertwined it with other fi-bres such as Big Sagebrush bark, Indian Hemp and White Clematis bark. They often coloured this twine with dyes made from Oregon-grape bark, Saskatoon berries and other plants. The Okanagan,

Secwepemc and Nlaka'pamux sometimes made Soapberry beaters from it by tying a bunch of shredded bark onto a short handle. It was said that in the old trading days in Okanagan country, three 12-cm-thick bundles of prepared Silverberry bark were worth a blanket. The Nlaka'pamux, Okanagan, Ktunaxa and the Blackfoot of Alberta used the large, striped seeds as beads for necklaces and for decorating clothing; they called them "buffalo beads". The Blackfoot rubbed them with grease and often interspersed them with juniper seeds.

A Silverberry seed necklace, and a Soapberry beater made with Silverberry bark by Mabel Joe of Merritt.

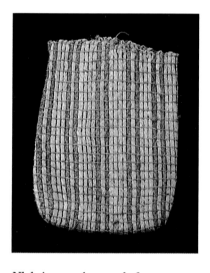

Nlaka'pamux bag made from Silverberry bark and Indian Hemp fibre, and dyed with Stawberry Blite berries. (RBCM 2653)

Arbutus
(Heather Family)

Arbutus menziesii
(Ericaceae)

Other Names: Pacific Madrone, Madrona.

Botanical Description

Arbutus is a branching broad-leaved evergreen tree, often twisted and gnarled, 6 to 30 metres tall. The bark is smooth and satiny, bright green when young, and maturing to a deep brownish-red; it is shed annually from the upper trunk and branches in large, paperlike sheets. The leaves are leathery, glossy, deep green and oval to elliptical; they are 7 to 15 cm long, and smooth-edged or finely toothed. They remain on the trees throughout the winter, turning yellow and dropping the following summer after the next season's leaves are completely developed. The flowers, which mature in spring, are white and urn-shaped, growing in dense sprays. The berries, up to 1 cm across, are bright red-orange, with a textured surface; they often remain on the trees well into winter.

Habitat: open moderately dry woods and rocky bluffs at lower elevations.

Distribution in British Columbia: on southeastern Vancouver Island, the Gulf Islands and the adjacent mainland west of the coastal mountains from Bute Inlet southward.

Aboriginal Use

Arbutus wood is too hard and brittle, and when drying, cracks too easily to be of much value for carving. Yet the Saanich once used the young branches to make spoons and gambling sticks. Recently, the Sechelt used the wood to construct the sterns and keels of small boats because it is durable under water. The Saanich placed the bark in steaming pits with Blue Camas bulbs to colour them pink. The Saanich and Cowichan boiled the bark as a tanning agent for paddles and fish hooks. Long ago, the Saanich strung the berries together to make necklaces.

Coastal Black Gooseberry and other wild gooseberries (Gooseberry Family)

Ribes divaricatum
Ribes species
(Grossulariaceae)

Botanical Description
Coastal Black Gooseberry is a stout, spreading deciduous shrub usually under 2 metres tall, with smooth, greyish bark and one to three sharp spines at each stem node. The leaves are small and shaped like maple leaves, with three to five main lobes. The flowers droop, and have deep-red sepals, white petals and protruding stamens. The fruits are smooth and purplish-black when ripe. There are several other gooseberry species in the province, all with the same general characteristics: short, bushy shrubs with spiny branches, palmately lobed leaves, and dark, drooping berries. One species, White-stemmed Gooseberry, is

very similar, differing only in a few minor features, such as the length and colour of the calyx lobes, which are slightly shorter and paler.

Habitat: Coastal Black Gooseberry grows in open woods, moist hillsides, clearings and shorelines; other gooseberries grow in a variety of habitats.

Distribution in British Columbia: Coastal Black Gooseberry occurs on southeastern Vancouver Island, the Gulf Islands and the Lower Mainland west of the Cascade Mountains, extending northward up the coast and eastward up some of the major river valleys; White-stemmed Gooseberry occurs on the eastern slopes of the Cascades; and other gooseberries are found throughout the province.

Aboriginal Use
Most aboriginal people do not distinguish among the different species of gooseberries, especially those with similar spine characteristics. The Saanich and Cowichan of Vancouver Island used gooseberry thorns as probes for opening boils, for removing slivers, and for tattooing. The Cowichan boiled the roots with Red-cedar roots and wild-rose roots and wove them together into rope, which they used to make reef nets.

The Nuxalk hollowed out the stems of Coastal Black Gooseberry to make pipe stems, with or without a bowl of elderberry wood. In the interior, the Secwepemc carved combs from the wood of one species of gooseberry and the Ktunaxa made fish hooks from gooseberry thorns, probably those of Mountain Gooseberry.

Mock-orange	*Philadelphus lewisii*
(Hydrangea Family)	**(Hydrangeaceae)**

Other Names: Syringa, Wild Orange, Soap Plant, "White Maple".

Botanical Description

Mock-orange is a handsome deciduous shrub up to 5 metres tall. It has light brown, shredding bark and pale green, lance-shaped to elliptical, pointed, coarsely toothed leaves usually 3 to 7 cm long. The flowers are white with yellow centres, four-petalled, and showy, in dense terminal clusters. In fragrance and appearance they resemble orange blossoms, hence the common name. The fruits are oval, pointed capsules, dark brown and hard when mature.

Habitat: shaded to open woods, gullies, talus slopes and rocky hillsides from sea level to subalpine elevations.

Distribution in British Columbia: in the southern part of the province, both on the coast and in the interior; the interior form is generally shorter and bushier, with flowers that are less fragrant than those of the coastal form.

Aboriginal Use

Mock-orange wood is strong and "hard as a bone", never cracking or warping when properly prepared. It was widely used for making implements, especially by the Interior Salish. The Okanagan used it for spear shafts, bows and arrows, digging sticks, "sliding" snowshoes, and clubs, as well as for pipe stems and cradle hoops. The Secwepemc used it to make digging sticks, combs, imitation breastbone decorations, fish spears and bear-paw type snowshoes. They sometimes obtained the wood from neighbouring Stl'atl'imx groups. The Stl'atl'imx used it mainly to make digging sticks, but Sam Mitchell of the Xaxl'ip Nation described how he made long knitting needles for his wife from an inch-thick Mock-orange stem. He split the stem in quarters, carved and sanded the pieces, fire-hardened the points, then polished them. Children of the Lower Stl'atl'imx made blowguns from the larger stems by hollowing out the pithy inner core. The Nlaka'pamux used the wood for making combs and as edging for birch-bark baskets and cradle hoods. The Ktunaxa made huckleberry-picking combs from it, and the Flathead of Montana made combs, pipe stems, bows and arrow shafts. On the coast, the Saanich sometimes used Mock-orange wood for bows and arrows, and the Lummi of Washington (also a Straits Salish group) made combs, netting shuttles and knitting needles from it. The Cowlitz of Washington made combs from it, and the Skagit used it for arrow shafts.

Mock-orange leaves foam into a lather when bruised and rubbed with the hands. The Okanagan, Secwepemc and Stl'atl'imx of British Columbia, and the Cowlitz and Snohomish of Washington used this foam as a soap for cleansing the skin; the Washington groups also used crushed Mock-orange flowers. The Secwepemc also obtained a lather from the bark by soaking it in warm water; they used the leaves for washing clothing.

To the Okanagan, the blooming of Mock-orange was a traditional indicator that the groundhogs (marmots) were fat and ready to be hunted.

Fireweed
(Evening Primrose Family)

Epilobium angustifolium
(Onagraceae)

Other Names: Willow Herb, Blooming Sally.

Botanical Description

Fireweed is a tall, smooth-stemmed herbaceous perennial, with spreading rhizomes
and alternate, smooth-edged, lance-shaped
leaves resembling those of narrow-leaved
willows. The flowers are red-purple with four
petals; they grow in long terminal clusters
and are very showy. They bloom throughout
the summer in sequence from bottom to top.
The seed capsules are long and narrow, splitting longitudinally on all four sides to reveal
rows of small parachuted seeds. The seeds
travel on the wind for long distances.

Habitat: open clearings, logged areas, burns and roadsides, in extensive patches. In summer the flowers often colour entire hillsides purple.

Distribution in British Columbia: widespread throughout the
province.

Aboriginal Use

The Haida made cordage from the outer stem-fibres of Fireweed, discarded when they ate the inner tissue. They peeled off the "skin" of the
stem and dried it. Later, they soaked it in water and twisted or spun it
into twine, which they used especially for making fishnets. Harlan
Smith (1997) noted that Gitxsan people also made cordage and fishnets
from Fireweed stems, gathered in June and July, before the plants flow-
ered and the stems became too hard; but they considered Fireweed
cordage inferior. Gitxsan people also used Fireweed string for weaving
pack straps. Sometimes they spun the fibre with Mountain Goat wool.

Several Coast Salish groups used the cottony seed fluff of Fireweed
mixed with other materials for weaving and padding. The Saanich and
other Vancouver Island groups and the Squamish of the mainland
added it to the wool of small domesticated dogs and wove the mixture
into blankets and clothing. According to one source, only higher-class

women of the Saanich group made clothing with dog's wool and Fireweed cotton. They first beat the dog's wool with diatomaceous earth, presumably to clean it. The Squamish also wove the Fireweed cotton with Mountain Goat wool. The Saanich stuffed mattresses with a mixture of Fireweed fluff and duck feathers. In Washington, most Puget Sound groups also used Fireweed seed fluff for weaving, mixing it with Mountain Goat wool. The Quinault and Skokomish used it mixed with duck feathers to make blankets.

The Carrier sometimes used Fireweed leaves to cover baskets full of berries. The Haisla used the leaves to wipe fish slime from their hands. The Blackfoot of Alberta rubbed the flowers on rawhide thongs and mittens to waterproof them. They applied the powdered inner stem tissue to the hands and face in the winter to protect them from the cold.

White Clematis (Buttercup Family)

Clematis ligusticifolia (Ranunculaceae)

Other Names: White Virgin's Bower, Pipestems, Traveller's Joy.

Botanical Description
White Clematis is a woody climbing vine with stems up to 20 metres long, often forming dense clumps along fences and thickets. The bark is greyish-brown and stringy. The leaves are pinnately compound, with five to seven oval, pointed, coarsely toothed leaflets. Male and female flowers grow on separate vines; both types are creamy white and grow in showy clusters. They bloom from May to August; often flowers and fruits appear simultaneously on the same plant. Each fruit – there are several per flower – bears a long, white plume that aids in dispersal. The plumes form large clouds of soft, silvery fluff, even more conspicuous than the flower clusters. They remain on the vines late into the fall.

Habitat: roadsides and creek bottoms, climbing on fences, shrubs and trees, or trailing over the ground, often forming dense thickets.

Distribution in British Columbia: common throughout the dry southern interior and also found on southern Vancouver Island and the Gulf Islands.

Aboriginal Use

The Okanagan and Nlaka'pamux stripped off the stringy bark fibre of White Clematis and wove bags, mats, capes and other garments from it. They often interwove or twined it with other plant materials, such as Indian Hemp or Silverberry bark. It was said that a good clematis bag could hold weights of 35 to 50 kilograms. In the old days, the Secwepemc used the vines, still growing, to straighten and strengthen implement handles. They placed the handle into the centre of a clematis plant, twining the stems around it, and left it there for up to two years, allowing the clematis to entwine it completely. When they disentangled it, they could use it as the handle of a tomahawk or some other implement. This treatment was said to make the handle very strong and keep it from warping.

The Okanagan, Flathead of Montana and others rubbed the leaves together to form a lather, then used this as a general soap and hair shampoo. The Flathead used the leaves of the Blue Clematis for the same purpose. A Ktunaxa consultant noted that the fluffy seed heads were used as nesting material by birds and "rock rabbits" (pikas).

Saskatoon Berry   *Amelanchier alnifolia*
(Rose Family)   (Rosaceae)

Other Names: Service Berry, June Berry, Shad-bush.

Botanical Description

Saskatoon Berry is a highly variable, deciduous shrub, 1 to 7 metres tall, with smooth reddish to grey bark. It has numerous round to oval leaves that are bluish-green and usually sharply toothed around the top half. The flowers, which bloom in April and May, are white and showy, with five elongated petals; they are crowded in drooping to erect clusters.

The flowers often cover the bushes – especially in the interior – creating a spectacular sight in areas where the plants are numerous. The berries, when ripe, are reddish-purple to dark blue, and sometimes seedy – the size, texture and taste vary considerably. Botanists distinguish three varieties in the province – var. *alnifolia,* var. *semiintegrifolia* and var. *cusickii* – but Interior Salish peoples designate up to nine types. Clearly, further taxonomic research on this highly complex species is warranted. This species has good potential as a garden shrub.

Habitat: dry woods and open hillsides in well-drained soil.

Distribution in British Columbia: common and widespread throughout the province, but most prolific in the dry woods and open slopes of the southern interior.

Aboriginal Use

Saskatoon Berry wood is hard, straight-grained and tough. It can be rendered even harder by heating it over a fire and is easily moulded while still hot. Its most important use was for making arrows. All of the Interior Salish groups – the Okanagan, Nlaka'pamux, Secwepemc and Stl'atl'imx – as well as the Ktunaxa, Carrier, Gitxsan, Straits Salish, Upper Sto:lo and the Flathead of Montana, used it for this purpose. In most of these cultures it was the major arrow-making material.

To make an arrow, a man chose a thin straight branch and stripped off the leaves and twigs. Stl'atl'imx men (and perhaps those in other groups) chewed the branch thoroughly to loosen the bark and break the grain of the wood to prevent it from curling or warping later. After removing the bark, the arrow maker fire-hardened the wood. He feathered one end and on the other fastened a bone, stone or metal point, or simply sharpened it. To finish it, he polished the surface with horsetail stems, and (usually) painted designs along the shaft.

Saskatoon Berry wood was also extensively used for making digging sticks, spear and harpoon shafts, implement handles, basket rims, and cradle frames. The Okanagan also made barbecue sticks and seed beaters from it, and the Secwepemc made barbecue sticks, rims for birchbark containers and thwarts for canoes. Both the Secwepemc and the

Stl'atl'imx placed a grid of green Saskatoon Berry sticks at the bottom of birch-bark cooking baskets to prevent them from being burned through by red-hot rocks. They also lined steaming pits with Saskatoon Berry twigs and used them as salmon spreaders for drying and cooking salmon. (The wood is said not to give a bitter flavour to the fish.) The Stl'atl'imx commonly constructed shelters from the branches for drying salmon and berries. The Carrier made slat armour and shields from the wood, covering them with animal hide, and wove mats from the branches on which to dry berries. On the coast, the Saanich used Saskatoon Berry wood to make herring rakes; and the Sto:lo, along the Fraser River, used it to make Eulachon rakes. They studded the ends of the rakes with rows of spikes, then swept the rake through the water to impale the fish.

The Nlaka'pamux and other Interior Salish peoples sometimes stained Silverberry bark (and other materials used in making bags) with mashed Saskatoon berries.

Red Hawthorn	*Crataegus columbiana*
Black Hawthorn	*Crataegus douglasii*
(Rose Family)	**(Rosaceae)**

Other Names: Red Thornberry, Red Haw, Columbia Hawthorn (*C. columbiana*); Black Thornberry, Black Haw, Douglas Hawthorn (*C. douglasii*).

Botanical Description
Hawthorns are tough, bushy deciduous shrubs or small trees. Red Hawthorn seldom grows more than 3 metres tall, while Black Hawthorn may reach 4 metres or more. Both species are armed with sharp thorns: those of Red Hawthorn are slender and 4 to 7 cm long, and those of Black Hawthorn are stouter and 1 to 3 cm long. The leaves are thick, dark

Red Hawthorn.

Black Hawthorn.

green and shiny, roughly oval, and coarsely toothed. Red Hawthorn leaves are pointed, while those of Black Hawthorn are usually blunter, though sharply toothed around the tip. The flowers, which bloom in April and May, are white and showy, growing in flat or rounded clusters. The berries of both species hang in clusters: Red Hawthorn berries are bright red, and those of Black Hawthorn are shiny black.

Habitat: well-drained sites near water, from gullies and stream courses to meadows and hillsides.

Distribution in British Columbia: Red Hawthorn is found in the southern interior from Prince George southward through the Okanagan and Kootenay valleys, and also in the Peace River District; Black Hawthorn is common throughout the province south of latitude 55°N.

Aboriginal Use
The Nlaka'pamux and Okanagan used the spines of both hawthorns as needles for probing boils and skin ulcers, and for piercing ears. The Okanagan used Red Hawthorn spines as pins in a ball-and-pin game (the ball was made of Tule stem) and the Nlaka'pamux made fish hooks

with them. The Stl'atl'imx and Gitxsan made trout hooks from Black Hawthorn spines.

People seldom used Red Hawthorn wood, but many used the tough, hard wood of Black Hawthorn: the Okanagan, Stl'atl'imx, Secwepemc and the Blackfoot of Alberta used it to make digging sticks; the Okanagan and Stl'atl'imx for clubs; the Secwepemc for axe handles, dipnet handles and double-tree yokes for horse-drawn wagons; and the Carrier and Gitxsan to make axe and adze

Black Hawthorn digging stick made by Secwepemc elder Mary Thomas.

handles. The Cowichan of Vancouver Island burned the leaves, inner bark and new shoots of the Black Hawthorn, mixed the ash with grease or ochre, and used it as a black face paint for winter dances.

Oceanspray   *Holodiscus discolor*
(Rose Family)   (Rosaceae)

Other Names: Arrow-wood, "Ironwood", Rock Spiraea.

Botanical Description
Oceanspray is an erect, deciduous shrub that grows 1 to 3 metres tall. Its branches are straight and slender, the young ones ribbed. The bark is reddish-brown when young, and grey on mature stems. The leaves are usually 4 to 10 cm long, bright green on the upper surface, paler beneath and more-or-less oval. The leaves have wedge-shaped bases and 15 to 25 shallow lobes, the toothed segments decreasing in size towards the tip. The flowers are small and cream-coloured, in large pyramidal clusters at the ends of the twigs – they make a conspicuous showing during the blooming season in mid summer. The fruiting clusters are greyish-brown and wispy, remaining on the bushes over the winter.

Habitat: dry rocky slopes and open woods from sea level to moderate elevations in the mountains.

Distribution in British Columbia: in the southern part of the province, both on the coast and in the interior; especially common on the rocky bluffs of southeastern Vancouver Island and the Gulf Islands.

Aboriginal Use
Most aboriginal people call this shrub "Ironwood" after the wood's hardness and strength. Like that of Saskatoon Berry, Oceanspray wood

can be made even harder by heating it over a fire. It was used by virtually all of the southern peoples – the Straits Salish, Halkomelem, Squamish, Sechelt, Kwakwaka'wakw, Stl'atl'imx, Nlaka'pamux, Okanagan and Ktunaxa – for making digging sticks, and by most of these groups for making spear and harpoon shafts, bows, and arrows. Some peoples, such as the Okanagan, bound their bows at the haft with deer sinew for added strength. Most people polished the wood with horsetail stems.

Some groups also had other uses for Oceanspray wood. The Okanagan used it to make tipi pins, gambling sticks, spear prongs, fish clubs, drum hoops, and hoops for baby cradles. The Nlaka'pamux made armour from it. The Lower Stl'atl'imx made hoops for attaching sinker stones to nets, cross-pieces for Douglas-fir dipnet hoops, and the upper sections of harpoon shafts and gaff handles. The Saanich and Cowichan made salmon barbecuing sticks, cambium scrapers, halibut hooks, Cattail mat needles and, recently, knitting needles. The Squamish made spear prongs, fish hooks, gaff sockets, harpoon shafts, clam-drying sticks, stiffeners for cedar-bark canoe bailers and mat needles. And the Sechelt made fire pokers, drum frames and clam-drying sticks. In western Washington, "Ironwood" was widely used to make roasting tongs, digging sticks and spear shafts. The Squaxin even made canoe paddles from it. Apparently, neither the Nuxalk nor the Secwepemc used the wood, although the bush does grow in their territories.

Wild Crabapple (Rose Family)

Malus fusca (Rosaceae)

Other Names: Pacific Crabapple, Western Crabapple, Oregon Crabapple.

Botanical Description
Wild Crabapple is a small, straggly tree, sometimes shrublike, 3 to 8 metres tall, with rough, grey bark on the trunk. The leaves, deep green and 4 to 10 cm long, are similar in shape to orchard apple leaves, but often have a prominent, pointed lobe along one or both edges. The flowers are white to pinkish and smaller than orchard apple blossoms;

they grow in flat to rounded clusters of 5 to 12. The crabapples hang in long-stemmed clusters. They are small, elongated, yellow to purplish-red when ripe, and very tart; after a frost they turn brown and soft. This species is also known as *Malus diversifolia*, *Pyrus fusca* and *P. diversifolia*.

Habitat: moist woods, stream banks, swamps and bogs, often in dense thickets.

Distribution in British Columbia: west of the coastal mountains from Vancouver Island to Alaska, up to 800 metres elevation.

Aboriginal Use
Wild Crabapple wood is hard and resilient. Coastal peoples such as the Kwakwaka'wakw, Straits Salish, Nuxalk and Halkomelem used it, as did the Gitxsan, to make implement handles, bows, wedges, sledge hammers, digging sticks and smaller items, such as gambling sticks, spoons, tongs and halibut hooks. The Kwakwaka'wakw treated the wood by scorching it over an open fire, then boiling it. The Saanich sometimes made fishing floats from it. The Squamish used it in recent times to make wedges and handles for axes and sledge hammers. The Nisga'a made pegs to hold their house boards in place from seasoned crabapple wood soaked in oil. In Washington, the Quileute made maul handles, seal-spear prongs and sea-bass lures from crabapple.

Bitter Cherry
(Rose Family)

Prunus emarginata
(Rosaceae)

Other Name: Wild Cherry, Bird Cherry.

Botanical Description

Bitter Cherry is a deciduous shrub or small tree up to 10 metres or more tall. Its bark is smooth and grey to shiny reddish-purple, and it peels off in horizontal strips. The leaves are short-stemmed, 3 to 8 cm long, el liptical to oval, round-tipped or bluntly pointed, and often finely

toothed. The flowers are white, in small clusters. The cherries are small and spherical, bright red to almost black, and exceedingly bitter.

Habitat: moist woods and clearings, often along watercourses; particularly abundant after a fire.

Distribution in British Columbia: abundant across the southern part of the province, especially on the coast and in the interior wet belt, from sea level to medium elevations in the mountains.

Aboriginal Use

Bitter Cherry bark was widely used by British Columbia First Peoples to decorate baskets and to wrap implements for protection and decoration. All the Coast and Interior Salish peoples, as well as the Carrier, Tsilhqot'in, Nuu-chah-nulth, Kwakwaka'wakw and others within the range of Bitter Cherry used the tough, shiny bark. They pulled it off the tree in thin, horizontal sheets, or for wrapping implements, cut it in a continuous spiral. They usually left it in its natural red colour or dyed it black. The most common method of dyeing was to soak it for several months in rich organic soil, such as in a swamp, or in manure, but it could also be coloured with vegetable dyes – the Nlaka'pamux used Flowering Dogwood or Grand Fir bark – or soaked in a can with rusty nails. Secwepemc elder Mary Thomas recalled that people sometimes dyed Bitter Cherry bark by placing it under the peeled-back bark of an alder root.

Secwepemc elder Mary Thomas peeling bark off a Bitter Cherry tree. When done properly this does not kill the tree.

Before they used the bark, the Nlaka'-pamux made it soft by pounding it or pulling it over the edge of a board or tree branch. Like other Salish groups, they used strips of dyed and natural bark, along with white and coloured grasses, to imbricate intricate patterns on their coiled split-root baskets, and also their mats and bags. They also used cherry bark to wrap splints for broken limbs and to reinforce suspension bridges, and they sometimes twisted it into twine. The Okanagan, Secwepemc and Stl'atl'imx used spirally cut strips of the bark to wrap around the hafts of bows as a hand grip and to add strength. The Okanagan glued the bark in place with fish slime mixed with ochre paint, and sealed the outside with tree pitch. They also used strips of the bark to decorate bows, tomahawk handles and pipe stems.

Bitter Cherry bark is waterproof, resistant to decay and makes a smooth union. These properties made it ideal for covering the joints of underwater implements, such as harpoons, dipnets, gaffs and fish spears. The Nuu-chah-nulth, Kwakwaka'wakw, Stl'atl'imx, Nlaka'-pamux and all the Coast Salish peoples used it in making such implements. Some of these peoples also used it to bind arrowheads onto shafts. They glued on the bark, then tied and sealed it with pitch, often applying several layers. The Lower Stl'atl'imx used Lodgepole Pine pitch boiled with bear grease as a glue and sealant. They made strings of twisted cherry bark to tie prongs onto harpoon shafts and as a binding for gaffs. The Comox wove fishing weirs with Bitter Cherry bark twine. The Kwakwaka'wakw covered wound dressings with the bark, sticking it onto the skin with pitch. The Squamish made cherry bark deer calls. In areas where the tree did not occur naturally, people often imported the bark from neighbouring locations; for example, the Manhousaht Nuu-chah-nulth obtained it from the Gold River area on Vancouver Island.

The Saanich, and perhaps other groups, considered Bitter Cherry wood an excellent fuel; they sometimes used it for the hearth and drill

Bitter Cherry bark – natural red and dyed black – with cured Reed Canary Grass, all prepared by Margaret Lester of Mount Currie for decorating coiled cedar-root baskets.

A coiled cedar-root basket, made by Margaret Lester, decorated with the prepared cherry bark and grass stems.

in making friction fires. Some contemporary artists carve small items out of Bitter Cherry wood.

The Okanagan occasionally carved darts (used in a throwing game) and other small items from the wood of the closely related Choke Cherry. The Nlaka'pamux made root-digger handles from Choke Cherry wood, and wove the shredded bark under the coils of the rims of baskets as ornamentation. The Secwepemc used Choke Cherry wood for salmon stretchers and mixed the cherries with bear grease to make a paint for colouring pictographs. The Gitxsan used the wood for adze handles and firewood, and they used the cherry pits as blowgun ammunition. The Blackfoot of Alberta used the wood for incense tongs and roasting skewers because it did not burn easily. The Secwepemc, Carrier and some other groups of the central interior also used the bark of another closely related wild cherry, Pin Cherry, in the same manner as they used Bitter Cherry bark.

Choke Cherry.

Antelope-brush
(Rose Family)

Purshia tridentata
(Rosaceae)

Other Names: Greasewood, Bitter Brush.

Botanical Description

Antelope-brush is an erect, gangly deciduous shrub, usually 1 to 2 metres tall, with dark grey to blackish bark. The leaves are small and wedge-shaped, deeply three-toothed at the tip, greenish on the upper surface, greyish beneath. The flowers are yellow and fairly small, but numerous and scattered along the branches, giving the bushes an overall yellowish appearance during flowering season, which is late spring. The fruits are small, dark, spindle-shaped and one-seeded.

Habitat: arid plains to dry open woods and slopes, usually with Big Sagebrush.

Distribution in British Columbia: restricted to the warm, dry areas in the southernmost part of the province – in the Okanagan Valley and the Rocky Mountain Trench.

Aboriginal Use

The pitchy quality of Antelope-brush makes it good for producing a hot fire quickly. The Okanagan used it for this purpose, breaking off the branches at ground level and tying them in bundles for burning. Antelope-brush bundles were especially useful during camping trips and in the winter. The Sanpoil-Nespelem Okanagan of Washington used the bark fibre for weaving bags, garments and soft baskets. It is said that deer like to browse this bush; you can smell it in their meat if they have been eating it. Okanagan children were warned not to play around Antelope-brush bushes because they have many wood ticks. The Okanagan considered the bush to be a type of sagebrush, although it is botanically unrelated. A.E. Chamberlain (1892) reported that the Ktunaxa obtained a reddish dye from the fruits.

Prickly Rose	*Rosa acicularis*
Dwarf Wild Rose	*Rosa gymnocarpa*
Nootka Rose	*Rosa nutkana*
Wood's Rose	*Rosa woodsii*
(Rose Family)	(Rosaceae)

Other Names: Bristly Rose (*R. acicularis*); Baldhip Rose (*R. gymno-carpa*); Common Wild Rose (*R. nutkana*); Prairie Rose (*R. woodsii*).

Botanical Description
Wild roses are erect shrubs with spiny or thorny stems and compound leaves, usually with five to seven toothed leaflets, similar to those of garden roses but smaller. The flowers are pink and have five petals, yel-

low centres and numerous stamens. The fruits (hips) are bright red-orange, consisting of a fleshy rind enclosing many whitish seeds. The rind is hard, but softens after the first frost; it tastes somewhat bland, but is high in Vitamin C. Prickly Rose has elongated fruits and numerous small spines on the stems and twigs. Dwarf Wild Rose has small flowers, small fruits without persisting sepals and usually densely bristled stems. Nootka Rose has large flowers and fruits, and one or two large, flattened thorns at each node, but no small spines. And Wood's Rose has smaller, straight thorns; smaller, more densely clustered flowers; and relatively small, round fruits.

Nootka Rose.

Habitat: Prickly Rose is found in open woods and damp meadows; Dwarf Wild Rose grows in moist shaded woods; Nootka Rose is common along roadsides and open woods, often forming dense thickets; and Wood's Rose is found in open woods and moist meadows, on prairies and creek sides.

Distribution in British Columbia: Prickly Rose occurs throughout the province east of the coastal mountains; Dwarf Wild Rose is found on both sides of the Cascade Mountains from about 52°N latitude southward; Nootka Rose is widespread along the coast and throughout the interior south of 56°N latitude; and Wood's Rose is common throughout the dry parts of the interior south of 56°N latitude and in the Peace River District.

Aboriginal Use

The wild roses were not as essential in aboriginal technology as some other plants, but they were put to a variety of uses in different regions. The Cowichan peeled and boiled the roots of the Nootka Rose and wove them together with boiled Coastal Black Gooseberry roots and Red-cedar roots to make reef nets. The Secwepemc made arrows of rose wood (probably Prickly Rose) and hollowed the stems to make pipe stems. The Nlaka'pamux used the wood of the Dwarf Wild Rose to make arrows, handles and baby-carrier hoops. The Okanagan used wild rose leaves to place over and under food in cooking baskets, steaming pits and pots to flavour the food and prevent it from burning. They sometimes made fishing lures by tying ant larvae onto a rose flower with horsehair. The Gitxsan used rose wood for arrow points. The Sechelt squeezed wild rose flowers to obtain a perfume. In pre-European times, such diverse groups as the Straits Salish of Vancouver Island and the Blackfoot of Alberta strung rose hips together to make necklaces.

Thimbleberry	*Rubus parviflorus*
(Rose Family)	**(Rosaceae)**

Botanical Description

Thimbleberry is an erect, many-stemmed shrub 1 to 2 metres tall. The bark is light brown, thin and shredded. The leaves are large and light green, resembling maple leaves with five pointed lobes; they have toothed edges and fine fuzz on both sides. The flowers are large and white, growing in terminal clusters. The berries turn from green to whitish to pink to bright red as they ripen. They are cup shaped and, when ripe, fall easily from their stems. Their flavour varies with locality and weather conditions, but ideally, they are sweet and tasty.

Habitat: open woods, clearings and roadsides, often forming dense thickets.

Distribution in British Columbia: widespread south of latitude 55°N; common along the coast north to Haida Gwaii.

Aboriginal Use

The Okanagan, Carrier and other peoples lined their steaming pits with the large, maple-like leaves of Thimbleberry, and many peoples used them to cover baskets of berries, to separate different kinds of berries in the same basket and as a surface for drying berries. Sometimes, too, people made a temporary berry-picking container by pinning the terminal lobes of a leaf together with a stick to produce a small cup. In Washington, the Quileute used the leaves to wrap cooked elderberries for storage, the Quinault used them, along with Skunk Cabbage leaves, to line elderberry preserving baskets, and the Cowlitz boiled the bark to make soap. The Gitxsan dried berries on Thimbleberry leaves; they also tied the leaves into a ball with willow bark twine, and folded them to make dishes and cornucopias for serving the berries. Some aboriginal peoples occasionally used Thimbleberries as a stain: the Blackfoot of Alberta dyed their tanned robes and coloured

Thimbleberry (left) and Wild Raspberry (below).

their arrow quivers with them. The berries of Wild Raspberry and Blackcap, both related to Thimbleberry, were also sometimes used as a stain for wood and other materials.

Salmonberry
(Rose Family)

Rubus spectabilis
(Rosaceae)

Botanical Description
Salmonberry is a tall, raspberrylike shrub with reddish-brown bark and numerous short prickles along the stems. The leaves, like those of raspberry, are compound, with two lateral leaflets and one larger terminal one. The flowers, usually solitary, bloom early in the spring before the leaves have fully expanded. They are pink and fairly showy. The berries resemble large raspberries and come in a range of colours, from salmon or gold to deep ruby red to almost black. Berries of different colour grow on different bushes, but can be found in the same locality.

Habitat: shaded swamps, damp woods and moist clearings along roads and shorelines, often forming large thickets.

Distribution in British Columbia: abundant along the coast from Vancouver Island to Alaska, west of the Coast Mountains.

Aboriginal Use
The Kwakwaka'wakw and Haida used the long, straight woody shoots of Salmonberry as spears in throwing games. The Kwakwaka'wakw and Nuu-chah-nulth sometimes used them for arrow shafts and the Haida used them to keep sheets of cedar-bark roofing flat by driving the sticks crosswise at intervals through the inner layers of the bark. The Squamish used short hollowed pieces of Salmonberry stem as the valves between harpoon heads and shafts and as the socket between the

hook and handle of a gaff. The Comox sometimes used a tube of the hollowed stem to add water to a covered steaming pit. The Makah of Washington made pipe stems from the hollowed stems and the Quileute used them to make plugs for seal-skin floats. Oweekeno people sometimes used Salmonberry leaves as an underlay for drying food. To Haisla and other coastal peoples the flowering of Salmonberry is an indicator that it is time to gather Red Laver.

Hardhack   *Spiraea douglasii*
(Rose Family)   (Rosaceae)

Other Names: Douglas Spiraea, Steeplebush, "Wild Lilac".

Botanical Description
Hardhack is an erect, wiry, deciduous shrub with reddish-brown bark. The oblong, elliptical to oval leaves are 4 to 10 cm long, dark green above, pale green beneath and toothed around the upper half. The flowers are small, pink to rose coloured, and numerous, in dense, elongated terminal clusters. The fruiting heads are deep brown, remaining on the bushes over the winter.

Habitat: stream banks, lake margins, bogs, swamps and damp meadows from sea level to subalpine elevations, often forming dense thickets.

Distribution in British Columbia: along the coast and in parts of the interior south from about 56°N latitude.

Aboriginal Use
The Nuu-chah-nulth used the wiry, branching twigs of Hardhack to make implements for gathering tubular dentalium shells from the mud-bottomed bays on the west coast of Vancouver Island. These shells, also called Money Tusk or Wampum, have been a valuable form of currency throughout northwestern North America since ancient times. The Nuu-chah-nulth were virtually the sole suppliers of dentalia in the North Pacific area. Long strings or wallet-sized packets of the

shells, all meticulously sorted by size, were traded from group to group as far east as the Great Plains. The implement for collecting them consisted of a broomlike bundle of Hardhack twigs or hardwood splints tied to a pole of Red-cedar or some other wood. A weighted board with a hole in the centre just big enough to slip part way down the bundle of twigs was placed over the pole, and more poles could be attached to lengthen the handle, depending on the depth of the dentalium beds. The shell gatherer took this implement out to the beds by canoe and lowered it into the water until it was nearly  touching the bottom. Then he pushed down hard into the mud, causing the twigs or splints to spread slightly, then close as the weighted board settled over them. With luck, the gatherer would find a few dentalia entrapped in the twigs when he pulled the implement up. In a rich area he could gather large quantities of the molluscs in a relatively short time. These he would clean and sort in preparation for a lucrative trade.

The Nuxalk used Hardhack branches to make hooks for drying and smoking salmon. The Vancouver Island Salish made blades, halibut hooks and cambium scrapers from the fire-hardened wood, and many peoples used the sticks to make salmon spreaders and roasting skewers.

Cottonwood:

Black Cottonwood *Populus balsamifera* ssp. *trichocarpa*
Balsam Poplar *Populus balsamifera* ssp. *balsamifera*
(Willow Family) (Salicaceae)

Other Names: poplar, Balm of Gilead (both); Northern Black Cottonwood (ssp. *trichocarpa*).

Botanical Description
Black Cottonwood and Balsam Poplar are rough-barked trees, up to 50 metres tall, with resinous, sweet-smelling spring buds and leaves. The leaves are long-stemmed and generally heart-shaped, triangular or more oval, with wedge-shaped bases. The leaf tips are sharply pointed and the margins finely toothed. In spring, the leaves are characteristically yellow-green and the buds are fragrant. The flowers are long, pendulant catkins, male and female being on separate trees. At fruiting time, the female catkins are covered with a soft, downy cottonlike substance, which is released with the seeds in mid summer, filling the air with bits of white fluff resembling snowflakes. Black Cottonwood has thick, leathery leaves and hairy fruit capsules; Balsam Poplar has thinner leaves and smooth fruit capsules. They often hybridize, but some botanists treat them as distinct species: *Populus trichocarpa* and *Populus balsamifera*.

Black Cottonwood.

Habitat: watercourses, gullies and floodplains; both are able to withstand periodic flooding.

Distribution in British Columbia: Black Cottonwood is found at low and high elevations throughout the province, except in the extreme northeast; Balsam Poplar occurs across northern British Columbia from the Peace River to the Yukon. These trees intergrade where their ranges overlap; neither has occurred on Haida Gwaii until recently, when some Black Cottonwoods were planted.

Aboriginal Use

First Peoples seldom distinguished Black Cottonwood and Balsam Poplar, so they are discussed here as one, simply Cottonwood. The wood of this tree is light brown, soft, moderately strong, straight-grained, uniform in texture and easy to work, but it is not very durable. It was used by many aboriginal peoples to make dugout canoes, including the Okanagan, Secwepemc, Carrier, Tahltan, Nisga'a, Gitxsan, Dena'ina of Alaska, Upper Sto:lo and occasionally the Vancouver Island Salish. Cottonwood canoes were smaller and lighter than the Red-cedar dugouts commonly used on the coast. Canoe makers widened a Cottonwood dugout by filling the hull with water and adding red-hot rocks to heat it, then spreading the sides with stout sticks. Cottonwood canoes were not as durable as cedar dugouts, and tended to become waterlogged, but they served well on the lakes and rivers of the interior, where large cedars were not available.

The Nlaka'pamux and Okanagan made saddles from Cottonwood. The Okanagan used Cottonwood boards in cradles to flatten their children's heads, and they cut Cottonwood poles to make fishing weirs. The Nisga'a sometimes carved masks from the wood. The Dena'ina of Alaska made roofing tiles from hollowed out half-logs of Cottonwood.

Cottonwood was considered an excellent fuel. The Okanagan, Secwepemc and Gitxsan used it, fresh or partially rotten, for smoking buckskin. The Stl'atl'imx and Tahltan used it for smoking fish, but a Secwepemc consultant maintained that it was too strong and would make the fish bitter. The Ktunaxa used dead, dried Cottonwood to bank a fire overnight. The Nlaka'pamux, Upper Sto:lo, Nuxalk and others used lengths of Cottonwood root, dried for a long time, to make drills and hearths for friction fires. The Okanagan used dried Cottonwood tops for the drills and dead roots for the hearths.

The Okanagan and Secwepemc made a soap substitute from Cottonwood ashes – the Okanagan for cleaning buckskin clothing and washing the hair, and the Secwepemc for laundering clothes. The Secwepemc put the ashes in a can of water and let them sit overnight. The next day they poured off the upper layer of liquid into another container, strained and bottled it. A small quantity of the fluid, placed in the laundry was said to act like lye. The Nlaka'pamux used the inner part of the bark as soap. They stripped off the bark when it was young and green, scraped off and discarded the outer part, and dried the white inner portion. This they packaged into small fist-sized bundles that could be stored for use in winter. In the early part of the 20th century, people carried their own soap bundles with them wherever they went.

It was said that the Hudson's Bay Company mixed the inner bark of Cottonwood trees with tallow to make soap, and some Nlaka'pamux people used it for laundry soap during the Second World War.

The Okanagan, Stl'atl'imx and some Coast Salish groups made rectangular containers from the thick and corky bark of mature Cottonwood trees, which they stripped off in large sheets from standing trees; they used the containers, or "buckets", for carrying and storing food. The Okanagan and others also used the bark to line and cover underground food caches to protect them from burrowing rodents. The Nisga'a sometimes built temporary cabins from Cottonwood bark. They also split the roots and twisted them into ropes for binding and tying fish traps and house planks.

The Okanagan, Stl'atl'imx, Secwepemc and others used the aromatic resin from the spring buds as a glue. They picked Cottonwood buds from the branches, or gathered the bud scales after they had fallen to the ground. To extract the resin, they heated the buds or scales and squeezed them. This glue was said to be the strongest cement known in the early days – stronger even than fish-slime glue. The Stl'atl'imx and Secwepemc used it to stick down sinew to bind feathers onto arrow shafts, and the Okanagan used it for gluing on arrowheads, spearheads and fish hooks, and for sealing the cracks in birch-bark canoes. The Oweekeno used it to fasten duck feathers onto cedar hoops for decoration. The Vancouver Island Salish also used this glue as waterproofing for baskets and boxes.

Some people heated the bud resin in oil and used it as a hair perfume and dressing, a skin salve, and a medicine for diaper rash. The Okanagan mixed the resin, which is a rich yellow colour, with various pigments – such as powdered alder bark, Western Larch pitch, Wolf Lichen or charcoal – to make paint. To prepare the resin for this purpose, they sliced the buds and boiled them in grease.

The Nlaka'pamux and Stl'atl'imx sometimes stuffed pillows and mattresses with the fluffy "cotton" from fruiting Cottonwood catkins, although it was hard to obtain in quantity. People throughout the province sometimes spun the cotton with Mountain Goat wool to make blankets and toques. Some peoples occasionally used Cottonwood branches as an underlay for cutting fish.

Trembling Aspen
(Willow Family)

Populus tremuloides
(Salicaceae)

Other Names: Aspen Poplar, White Poplar.

Botanical Description

Trembling Aspen is a slender, medium-sized, deciduous tree that grows up to 25 metres tall, with few branches on the lower trunk. The bark is smooth and light green to whitish. The leaves are broadly oval – rounded at the base and pointed at the tip – and finely toothed. The leafstalks are vertically flat near the blade, so that the leaves flutter in the slightest breeze. The leaves turn golden in autumn, making a stand of the trees a spectacular sight. The flowers hang in catkins, male and female on separate trees, and they mature before the leaves expand in the spring. Aspens also reproduce vegetatively by the growth of suckers from the root system.

Habitat: open meadows and mixed coniferous forests.

Distribution in British Columbia: abundant in the central, northeastern and southern interior, from near sea level to subalpine elevations, where it forms extensive stands; also occurs sporadically along the coast and on Vancouver Island, but not on Haida Gwaii.

Aboriginal Use

The wood of Trembling Aspen, nearly white in colour, is soft, brittle and not very durable, but it seasons well and is easy to work. The Upper Nlaka'pamux sometimes used it to make small dugout canoes, although it was said to be heavier than Red-cedar. The Okanagan scraped deer hides on Aspen logs. They preferred to cut the logs in early May, because the bark peels off easily at this time. Secwepemc and Blackfoot boys made whistles from the branches, and the Secwepemc used the thin trunks as tent poles and to make drying racks for fish and deer meat, but they said that the poles would rot after only a couple of years. The Ktunaxa made hide-covered saddles from Aspen wood. The Secwepemc and Blackfoot and probably most other interior groups used the wood as a fuel. The Carrier used rotten Aspen wood for lining babies' cradles; it is soft and absorbent and was said to make good diaper material.

The Nlaka'pamux made a cleansing solution for washing guns, traps and buckskins by boiling Aspen branches in water. Hunters and others washed themselves in this liquid to clean their skin and eliminate human odour. The Okanagan used Aspen and Cottonwood as a weather indicator: if the leaves began to shimmer when there was no perceptible wind, it would soon become stormy. The Dena'ina of Alaska used Trembling Aspen branches as steam bath switches and also as Beaver bait.

Willows	*Salix* species
(Willow Family)	**(Salicaceae)**

Botanical Description

Willows are deciduous shrubs or trees with rounded to elongated leaves. The flowers grow in catkins, male and female on separate plants. Most willows are wind pollinated and their fluffy fruits are carried on the wind. There are nearly 50 species of willows in British Columbia. They vary widely in size and habit, bark colour and texture, leaf shape, hairiness, and other characteristics. Many species hybridize freely. Some of the species used by First Peoples are:

Pacific Willow (*Salix lucida* ssp. *lasiandra*): a slender, narrow-leaved tree, up to 12 metres tall.

Sandbar Willow (*S. exigua*): a spreading shrub up to 3 metres high with long, slender, erect stems, smooth pinkish-brown bark, and long, narrow leaves that are greyish-green and usually lightly toothed at the margins; also called "Pink-barked Willow", "Silver Willow" or Coyote Willow.

Scouler's Willow (*S. scouleriana*): a large shrub or small tree with smooth, grey bark and oblong "mouse-ear" leaves that taper at the base; the undersides of the leaves are often covered with rust-coloured hairs.

Pacific Willow (left) and Hooker's Willow (above).

Hooker's Willow (*S. hookeriana*): a large shrub or small tree with oval, woolly, smooth-edged leaves.

Mackenzie Willow (*S. prolixa*): a tall shrub with yellow or brownish bark and finely toothed, lance-shaped leaves.

Bebb's Willow (*S. bebbiana*): a tall shrub or small tree with grey-brown bark and oblong oval to lance-shaped leaves that are whitish and usually thinly hairy beneath.

Many other species were used, but most were not distinguished as separate types by First Peoples or the authors of published sources – these are treated only in a general context. Where the species used is known, it is mentioned.

Habitat: most willows grow in moist areas – in swamps and along watercourses; all of the species mentioned grow in wet ground. Sandbar Willow forms dense colonies on sand and gravel bars along rivers and streams; Scouler's Willow grows on upland slopes, and in moderately dry woods and moist gullies; Bebb's Willow also grows in upland sites.

Distribution in British Columbia: willows are found throughout the province, from sea level to alpine elevations. Pacific Willow grows throughout the province; Sandbar Willow is widespread in the interior east of the coastal mountain ranges; Scouler's Willow occurs through-out the province, but is especially common along the southern coast; Hooker's Willow is restricted to the coastal regions, mainly in the south; Mackenzie Willow grows throughout the province, but especially in the east; and Bebb's Willow occurs throughout the interior east of the coastal mountains.

Aboriginal Use

The Fraser River Stl'atl'imx called Pacific Willow the "match plant". They dried the wood and used it for both the hearth and the drill in making friction fires, using dry grass and Big Sagebrush bark as tinder. The Cowlitz of Washington also made fire-making drills from Pacific Willow wood. Some Stl'atl'imx groups used a willow fire to burn di-atomaceous earth into a fine white powder for treating wool. The Okanagan used the wood of Bebb's Willow, Scouler's Willow and others to make hide stretchers, barbecue sticks and fish traps, and used the twigs for smoking hides when they desired a white colouring. They used willow twigs daily to clean their teeth. The Secwepemc used the wood of Bebb's and Scouler's willows for smoking salmon, drying meat and fish, and for making barbecue sticks and fishing weirs. They used rotten willow roots as punk or tinder, which could be ignited and carried while travelling. The Ktunaxa considered willow wood an ex-cellent fuel for smoking meat. The Flathead of Montana and the Blackfoot of Alberta made sweat-houses from willow branches, the Blackfoot specifically from Sandbar Willow. The Tahltan used willow for smoking fish and for making gambling sticks and snowshoes. The Gitxsan used willow for firewood and willow roots for fire-making drills and hearths. The Sekani made small bows from Sandbar Willow. The Carrier made berry drying racks, pack boards and snowshoe cross-sticks from willow wood. For snowshoes, willow was said to be light but not very durable.

Stl'atl'imx elder Sam Mitchell of Xaxl'ep (Fountain) twists the supple branches of "rope plant" (Sandbar Willow) to make rope.

The Haida made spoons and other small articles and frames for summer houses from Scouler's Willow. The Kwakwaka'wakw made willow knitting needles and game hoops. The Vancouver Island Salish sometimes made bows from willow wood. The Haisla made mallet heads from Pacific Willow. The Quileute of Washington used the branches of young Hooker's Willow as poles for fish weirs, saying that they would take root wherever they were "planted" in the river.

The supple, flexible nature of willow branches and bark, especially of some species, made them extremely useful as cordage, for lines, nets and ropes, and for binding and tying. The Straits Salish and Halkomelem peeled the bark of Hooker's Willow and other species in May or June, removed the outer part, split the inner tissue into thin strands, and twisted these together into a long rope. They used the rope to make fishing lines and various types of nets, including gill nets, reef nets, purse nets, bag nets and even duck nets. They also used the bark for basket decoration.

The Snohomish and Quinault of Washington made tumplines, slings and harpoon lines from Hooker's Willow bark. The Clallam made string from the bark of the Sitka Willow and the Chehalis used the inner part of Pacific Willow to make a two-ply string. The Lower Stl'atl'imx also made twine, for tying fish traps, from Pacific Willow bark, and made strong ropes by braiding the bark and twigs together. They used these as anchor lines and for attaching cedar floats to nets. Some Stl'atl'imx people, reportedly only the poorer ones, shredded the inner bark of willow and wove it into robes, skirts, bodices, aprons and socks. The Stl'atl'imx called Sandbar Willow "rope plant" and used it more than any other willow for making rope. The branches, bark and all, when twisted into rope while still green, remained pliable even when dry.

The Nlaka'pamux and Stl'atl'imx wove bags, mats, capes, aprons and blankets from willow bark; but the Nlaka'pamux, according to James Teit (Steedman 1930), only used the bark from dead trees. The Nlaka'pamux used Sandbar Willow withes in construction, and wove the bark into saddle blankets. The Okanagan twisted the bark of Bebb's, Sandbar and other willows into cord for tying rafts and fish traps together and for weaving bags, dresses and skirts. In making a woman's skirt, they stretched the willow bark while still green, hung it up to dry for a day, then softened it by rubbing a stone scraper across it many times. They shredded the inner

Willow-bark dolls made by Mary Thomas.

bark into a cottonlike substance and used it for diapers, wound dressings and women's sanitary napkins. The Okanagan laid fish on Sandbar Willow branches, and used willow leaves for wrapping and serving fish.

The Secwepemc used withes of Scouler's Willow and others for sewing birch-bark canoes, and for making cradle rims and hoops. They used strips of inner bark to make headbands and to string edible roots for drying. They also used strips of bark for lashing and tying. The Carrier made fishnets of Sandbar Willow bark and the Tahltan used the bark and twigs for tying house walls together. The Haisla used withes of Pacific Willow and Scouler's Willow to string drying fish; they used the leafy branches to wipe fish and as a matting for draining and cutting salmon; they also made walking sticks from Scouler's Willow.

The Secwepemc made dolls from the inner bark of a willow (Mackenzie Willow and, possibly, Pacific Willow). Secwepemc elder Mary Thomas remembered her grandmother making her a new willow-bark doll every year. The Dena'ina of Alaska used the inner bark of large willows for fishnets, fish hangers and lashings; they made whistles and temporary snowshoes from willow withes; and they used the leafy branches for flooring and thatching temporary shelters. The Ktunaxa twisted the green bark of Sandbar Willow and other species to make lashing for rafts and other things. The Flathead of Montana used willow bark to make ropes and bridles, and to weave baskets.

Some groups, such as the Haida, used the spring catkins, or "pussy willows" of certain willows as decoration, at least in recent years. The Straits Salish sometimes used willow bark to make a grey dye for colouring Mountain Goat wool. Many peoples considered willows an excellent food for Beavers.

Stinging Nettle
(Nettle Family)

Urtica dioica
(Urticaceae)

Other Names: "Indian Spinach", Northwest Nettle.

Botanical Description

Stinging Nettle is a herbaceous perennial that grows 1 to 3 metres tall and has spreading rhizomes. The ragged-looking leaves grow on short stalks in opposite pairs along the stem. They are 7 to 15 cm long, rounded at the base, broadest below the middle and tapering to a sharp point. The edges are sharply toothed. The greenish, inconspicuous flowers are clustered in drooping bunches at the stem nodes. The leaves and stems are covered with stiff hairs that cause stinging and blistering when touched. Another name for *Urtica dioica* is *U. lyallii.*

Habitat: on the edges of clearings and old fields, and in damp roadside thickets and shaded woods, usually in large patches.

Distribution in British Columbia: common along the coast from Vancouver Island to Alaska, and in the southern and central interior.

Aboriginal Use

Stinging Nettle stems were an important source of fibre for most coastal peoples – the Straits Salish, Halkomelem, virtually all the western Washington groups, the Squamish, Kwakwaka'wakw, Nuu-chah-nulth, Nuxalk, Tsimshian, Haida and Tlingit – as well as the Nisga'a, Gitxsan, Carrier, Lower Stl'atl'imx and Lower Nlaka'pamux. The Interior Salish people preferred to use Indian Hemp fibre. People gathered Stinging Nettle stems in the fall, usually in October, after the plants had completely matured and were beginning to die. The Nuxalk harvested them when the first snow hit the mountain tops.

To prepare the fibre, a person (usually a woman) stripped off the leaves and dried the stems in the sun for several days. (Some Vancouver Island Salish also dried them over a fire.) Then she cracked off the brittle inner pith in short sections, separating the outer fibres. She

Stinging Nettle fibre.
(RBCM 13645WC)

gently pounded the fibres or worked them with her hands to remove the thin outer skin. The individual strands of fibre are whitish in colour. The weaver spun them on her bare thigh or with a wooden disc spindle, often made of Broad-leaved Maple wood, splicing the strands together by rolling or twisting them. Finally, she twisted the resulting thread into a two-ply or four-ply twine. This twine could be used for tying and binding and for making tumplines and pack straps, bowstrings, snares, harpoon and fishing lines, fishnets, duck nets, and even deer nets. In Nuxalk culture, the men made the fishnets, although as everywhere, the women prepared the twine. Nettle fishnets were often dyed brown with alder bark to make them invisible to fish.

The Saanich of Vancouver Island used to spin nettle fibre with bird down to make blankets and sleeping bags in the days before Mountain Goat wool could be easily obtained from the mainland. The Cowichan used nettle fibre for tattooing – they rubbed charcoal or another pigment on a nettle thread and ran it beneath the skin with a fine hardwood needle.

The Tlingit of Alaska made a red dye by boiling Stinging Nettle stems and leaves in urine. The Okanagan poisoned arrows by boiling the points in water with nettle roots. According to Squamish lore, the first Stinging Nettle shoots coming out of the ground in spring signalled that seals were giving birth to their young.

APPENDIX 1

Minor Plants in First Peoples' Technology

This section contains brief descriptions of flowering plants that had minor roles in the technologies of the First Peoples of British Columbia and adjacent areas. The plants are divided into the major groups of Monocotyledons and Dicotyledons, then alphabetically by scientific family names, and finally alphabetically by scientific species names.

MONOCOTYLEDONS

Sedge Family   Cyperaceae

Common Spike-rush   *Eleocharis palustris*
The Okanagan used Common Spike-rush stems for bedding and for sitting on in the sweat-house.

Rush Family   Juncaceae

Common Rush   *Juncus effusus*
The Comox and possibly some other peoples occasionally used Common Rush stems for weaving. The Quinault of Washington twined the stems into tumplines and intertwined them with Cattail to make string, and the Snuqualmi used Common Rush stems for tying things. (See photograph on next page.)

Common Rush.

Lily Family   Liliaceae

Nodding Onion   *Allium cernuum*
The Vancouver Island Salish rubbed Nodding Onion bulbs on their skin to repel mosquitoes and other insects.

Queenscup   *Clintonia uniflora*
The Nlaka'pamux mashed Queenscup berries and used them as a blue dye, but they needed a large quantity to make the dye effective.

Clasping Twisted Stalk   *Streptopus amplexifolius*
False Solomon's Seal   *Smilacina racemosa*
Star-flowered Solomon's Seal   *Smilacina stellata*
The Nlaka'pamux and Secwepemc used the roots or the whole plants of these species as a scent, tying them to the body or on the clothes or hair. If the Chinook Salmon catch was poor, a Lower Stl'atl'imx fisherman would rinse his gill net in a solution of boiled Clasping Twisted Stalk to bring good luck in fishing.

False Hellebore   *Veratrum viride*
James Teit (in Steedman 1930) reported that the Stl'atl'imx used the stem fibres of this poisonous plant to weave wallets, bags and pouches.

Death Camas *Zigadenus venenosus*

The Okanagan mashed the bulbs of Death Camas to make an arrow poison.

DICOTYLEDONS

Sumac Family Anacardiaceae

Smooth Sumac *Rhus glabra*

The Okanagan people know that when the leaves of Smooth Sumac change colour in the fall, the Sockeye Salmon are also turning red. The Nlaka'pamux experimented with the fruits as a source of red dye, but the results were not satisfactory.

Celery Family Apiaceae

Kneeling Angelica *Angelica genuflexa*

Ktunaxa, Gitxsan and Haisla children made whistles and blowguns from hollow Kneeling Angelica stems, but they had to be careful not to confuse them with those of the Douglas's Water-hemlock, which are very poisonous. Gitxsan boys used Choke Cherry pits as ammunition for their blowguns. The Gitxsan and others also used the hollow stems as drinking straws. The Nuxalk used the stems as breathing tubes when hiding under water in times of danger. Some Haisla people used the stems to collect and store spruce pitch for chewing. In Washington, Makah and Quileute children made whistles out of the related Water Parsley, but this is not recommended, because this plant is easily confused with Douglas's Water-hemlock, and may itself be poisonous. See the Warning under Cow Parsnip on page 133.

Douglas's Water-hemlock   *Cicuta douglasii*
The Okanagan used powdered Douglas's Water-hemlock root as an arrow poison. This plant is extremely toxic. See the Warning under Cow Parsnip on page 133.

Birthwort Family   Aristolochiaceae

Wild Ginger   *Asarum caudatum*
The Nlaka'pamux and Okanagan sometimes mixed Wild Ginger with sphagnum moss as a bedding for infants. The Lower Stl'atl'imx and others used Wild Ginger in their bath water. People everywhere valued Wild Ginger for its pleasant spicy scent.

Aster Family   Asteraceae

Yarrow   *Achillea millefolium*
The Haida used Yarrow stems to skewer Butter Clams for drying. Yarrow imparted a pleasant taste to the food, and people ate the dried clams right off the stems. The Okanagan, Secwepemc and others placed Yarrow leaves on hot coals to make a smudge for repelling mosquitos. The Flathead of Montana rubbed the flower heads in their armpits as a deodorant. Sometimes people rubbed the aromatic leaves directly on the skin as an insect repellent, but it has a reputation for causing spots on the skin.

Pearly Everlasting   *Anaphalis margaritacea*
The Nlaka'pamux stuffed pillows with the dried flower heads of Pearly Everlasting. Secwepemc women used dried, shredded leaves and flowers as sanitary napkins. The Gitxsan placed flowers on coffins and graves.

Grey Rabbitbrush   *Chrysothamnus nauseosus*
Most aboriginal people consider Grey Rabbitbrush a type of sagebrush, or closely related to it. The Okanagan sometimes used the pungent smelling branches for smoking hides. The Sanpoil-Nespelem Okanagan of Washington pulverized the leaves and twigs, and rubbed

Wild Ginger.

Grey Rabbitbrush.

them on horses to protect them from horseflies and gnats. The Secwepemc stuffed pillows and mattresses with the cottony fruiting heads.

Subalpine Daisy   *Erigeron peregrinus*
The Nlaka'pamux name for Subalpine Daisy means "star flower", according to James Teit (in Steedman 1930). They used the flower head as a basketry pattern, and said that it was easier to duplicate than other flowers.

Pineapple Weed   *Matricaria matricarioides*
Ktunaxa children strung Pineapple Weed flower heads for necklaces. Adults hung the plants in the house because of their nice smell, and stuffed pillows with them. The Flathead of Montana used Pineapple Weed as an insect repellent: they placed whole plants in alternating layers with meat or berries in skin bags to keep the bugs off, and sprinkled dried leaves over fresh meat and fruit.

Sweet Coltsfoot   *Petasites frigidus*
The Quinault of Washington used Sweet Coltsfoot leaves to cover berries in steaming pits. The Ditidaht name for this plant means "Elk's food", because Elks eat the leaves.

Arrow-leaved Coltsfoot   *Petasites sagittatus*
Secwepemc women in the Salmon Arm area used to gather Arrow-leaved Coltsfoot leaves in quantity. They dried the leaves, rubbed them

Drying Arrowleaf Coltsfoot leaves.

in their hands and removed the large veins. Working the leaves in this way produced a fine cottony material that could be used as a sanitary napkin. Secwepemc elder Mary Thomas recalls her sister using it wrapped in cheese-cloth.

Canada Goldenrod   *Solidago canadensis*
Okanagan children used to play with Canada Goldenrod plants, pulling them up and using them as whips.

Honeysuckle Family   Caprifoliaceae

Highbush Cranberry   *Viburnum edule*
The Ktunaxa made pipe stems from the hollowed-out branches of Highbush Cranberry. The Dena'ina of Alaska made a red dye from the fruits. They used the branches as rims for birch-bark baskets and the leafy twigs as steam-bath switches. The Gitxsan made ropes from the twigs and used them to tie rafts together.

Boxwood Family   Celastraceae

False Box   *Pachistima myrsinites*
In recent years, Saanich women gathered False Box branches and sold them to local florists for use in floral decorations.

Goosefoot Family   Chenopodiaceae

Lamb's Quarters   *Chenopodium album*
The Blackfoot of Alberta obtained a green dye from the young shoots
of Lamb's Quarters.

Jerusalem Oak   *Chenopodium botrys*
Jerusalem Oak is a common Eurasian weed, but according to James
Teit (in Steedman 1930), the Nlaka'pamux used it in large quantities as
a scent. They wound it in necklaces, stuffed it in pillows, bags, pouches
and baskets, and tied it onto clothes and in hair or wore it in little bags
next to the skin.

Strawberry Blite   *Chenopodium capitatum*
The Nlaka'pamux, Tsilhqot'in,
Carrier and others crushed the
fruits of Strawberry Blite to make
a red stain to colour the face and
body, as well as clothes, skins,
wood, and fibres for weaving. (See
the photograph of the Silverberry
bark bag on page 170.) The Carrier sometimes coloured spruce
roots with Strawberry Blite.

Bindweed Family   Convolvulariaceae

Field Bindweed   *Convolvulus arvensis*
Field Bindweed is an introduced weed. In recent years, Okanagan
hunters twined the stems into a pack-rope for carrying birds and marmots home.

Heather Family   Ericaceae

Kinnikinnick   *Arctostaphylos uva-ursi*
The Stl'atl'imx sometimes used Kinnikinnick roots to make temporary
pipes. The Blackfoot of Alberta used the dried berries in their rattles
and strung them to make necklaces.

Salal   *Gaultheria shallon*
The Saanich and other Vancouver Island Salish groups placed Salal
branches in their steaming pits, over and under such foods as Blue
Camas bulbs. The Nuu-chah-nulth made a purple stain from the
berries. The Kwakwaka'wakw mixed them with Black Twinberry fruits
to intensify the colour of the stain. They and the Squamish used the
branches to whip Soapberries.

False Azalea   *Menziesia ferruginea*
The Quileute of Washington wove False Azalea twigs with cedar bark
to make a mat for sitting on in the bottom of a canoe. The Haida placed
the twigs in coffins.

Pinedrops   *Pterospora andromedea*
The Flathead of Montana boiled Pinedrops with Blue Clematis to make
a shampoo.

White Rhododendron   *Rhododendron albiflorum*
The Nlaka'pamux used White Rhododendron as a scent.

Black Huckleberry   *Vaccinium membranaceum*
The Okanagan, Stl'atl'imx and Tlingit of Alaska mashed Black
Huckleberries to obtain a purple dye for basket materials.

Blueberries and Huckleberries   *Vaccinium* species
Haida elders have reported that blueberry wood was used to make pegs
for fastening the joints of kerfed cedar-wood boxes. The Dena'ina of

Alaska and other peoples used blueberries as a purple dye for snowshoes, grass mats, birch-bark baskets, porcupine quills and skin. The Dena'ina also used other berries as dyes, including Lowbush Cranberry, Crowberry, Black Currant, Trailing Black Currant and Highbush Cranberry.

Huckleberries in a birch-bark basket.

Pea Family   Fabaceae

Timber Milk-vetch   *Astragalus miser*
The Okanagan used Timber Milk-vetch to wipe the turpentinelike juice from the inside of Lodgepole Pine bark when they harvested the cambium.

Beach Pea   *Lathyrus japonicus*
The Nuu-chah-nulth of Manhousat placed the sweet-smelling flowers of Beach Pea in small grass baskets to scent them so that they would sell better. Some Gitxsan people used a related species, Creamy Peavine, to wipe salmon or meat; this species is well known as good fodder for livestock.

Silky Lupine   *Lupinus sericeus*
and other lupines   *Lupinus* species
The Okanagan used lupine flowers as bedding and flooring in the sweat-house. They considered lupines the marmot's favourite food – lupines blooming in spring was a sign that the marmots were fat enough to eat.

Giant Vetch   *Vicia gigantea*
Saanich people used Giant Vetch to make a hair rinse, and fishermen
rubbed their hands on it before handling their gear. The Makah of
Washington used the leaves and vines to cover food while steaming.

Beech Family   Fagaceae

Garry Oak   *Quercus garryana*
The Cowlitz of Washington used Garry Oak wood to make combs and
digging sticks, and burned it as a fuel.

Gooseberry Family   Grossulariaceae

Stink Currant   *Ribes bracteosum*
The Quileute of Washington used the hollowed stems of Stink Currant
to inflate seal stomachs used as oil containers. They also used the leaves
to line and cover elderberry storage containers (made of Western
Hemlock bark).

Mint Family   Lamiaceae

Field Mint   *Mentha arvensis*
The Nlaka'pamux, the Secwepemc and the Flathead of Montana used
Field Mint as a scent. The Secwepemc placed it under pillows and kept
it around the house "just for the smell of it". One Secwepemc elder recalled that her grandmother washed her and her siblings in a mint solution when they were sprayed by skunks. The Flathead put the plants
in the corners of houses, on the floors of sweat-houses, and kept it in
suitcases and around the house to keep insects away. The Blackfoot of
Alberta boiled their traps in a mint solution to destroy the human scent.

Field Mint.   Wild Bergamot.

Wild Bergamot   *Monarda fistulosa*
The Ktunaxa and the Flathead of Montana dried and powdered Wild
Bergamot leaves and sprinkled them over fresh meat or fruit as an in-
sect repellent and preservative. Ktunaxa people also placed the leaves
on rocks in the sweat-house as a perfume. Secwepemc people used
Wild Bergamot as a smudge for repelling mosquitoes.

Coyote Mint   *Monardella odoratissima*
The Colville and Sanpoil-Nespelem Okanagan of Washington wiped
Coyote Mint on spears, harpoons, animal snares, arrows, fishing hooks
and fishing lines to clean them and remove the human scent.

Cooley's Hedge-nettle   *Stachys cooleyae*
The Makah and Quinault of Washington covered steaming sprouts
with Hedge Nettle plants. Nuu-chah-nulth fishermen wiped their hands
on this plant before handling their gear.

Phlox Family   Polemoniaceae

Scarlet Gilia   *Ipomopsis aggregata*
The Flathead of Montana dried Scarlet Gilia plants and put them in
bags to make sachets.

Buckwheat Family   Polygonaceae

Parsnip-flowered Buckwheat   *Eriogonum heracleoides*
and other buckwheats   *Eriogonum* **species**
Okanagan children played a game with Wild Buckwheat stems. They
broke off the main stem, leaving one side-branch attached, to make a
hook. Each child took a stem and they all hooked them together and
pulled. The first one to break his stick lost the game.

Primrose Family   Primulaceae

Pretty Shooting Star   *Dodecatheon pulchellum*
The Okanagan mashed Pretty Shooting Star flowers and smeared them
on arrows as a pink stain.

Parsnip-flowered Buckwheat: hooking
the stems.

Buttercup Family   Ranunculaceae

Long-headed Anemone   *Anemone multifida*
The Nlaka'pamux pounded the seed fluff of Long-headed Anemone into "flannel cloth" and used it for baby diapers. They made a strong decoction from the whole plant to kill lice and fleas.

Montana Larkspur   *Delphinium bicolor*
The Okanagan used the bright blue flowers of Montana Larkspur and other larkspurs as a blue dye for colouring arrows and other items. The Nlaka'pamux also used the flowers of a larkspur as a blue colouring, even for dyeing clothing, but they were reported by some to be of little value. The Blackfoot of Alberta mixed the flowers with water and used the solution to dye Porcupine quills. Recently, the Secwepemc mixed larkspur flowers with rose petals for the Corpus Christi procession in the church. They were scattered by small girls.

Sagebrush Buttercup   *Ranunculus glaberrimus*
The Nlaka'pamux washed Sagebrush Buttercup flowers or the whole plants and rubbed them on arrow points as a poison. Other species of buttercup could be used if this one was not available.

Western Meadowrue   *Thalictrum occidentale*
The Flathead of Montana pulverized the dried seeds of Western Meadowrue by chewing them, then they rubbed them on the hair and body as a perfume.

Buckthorn Family   Rhamnaceae

Redstem Ceanothus   *Ceanothus sanguineus*
The Okanagan used the wood of Redstem Ceanothus (also called Buckbrush) as a fuel for smoking deer meat if other woods were not available. The branches, especially the buds, were said to be an important food for deer in spring.

Snowbrush *Ceanothus velutinus*
The Secwepemc placed Snowbrush branches on a hot stove to fumigate a house. The smoke acted both as a disinfectant and an insect repellent.

Cascara *Rhamnus purshiana*
The Nuu-chah-nulth of Manhousat made chisel handles from Cascara wood. The Skagit of Washington boiled the bark to make a green dye for Mountain Goat wool.

Cascara.

Rose Family Rosaceae

Wild Strawberries *Fragaria* **species**
Stl'atl'imx girls used to wear headbands and belts of strawberry runners plaited together in three or four strands.

Three-flowered Avens *Geum triflorum*
The Blackfoot of Alberta crushed the ripe seeds of Three-flowered Avens and used them as a perfume.

Pacific Ninebark *Physocarpus capitatus*
The Cowichan of Vancouver Island recently made knitting needles from Pacific Ninebark wood, and the Nuu-chah-nulth used it to make children's bows and other small items.

Mallow Ninebark *Physocarpus malvaceus*
The Southern Okanagan of Washington sometimes made bows from Mallow Ninebark wood.

Common Silverweed   *Potentilla anserina*
The Blackfoot of Alberta used Common Silverweed (Cinquefoil) runners as ties for leggings and blankets.

Shrubby Cinquefoil   *Potentilla fruticosa*
The Blackfoot of Alberta used the dry, flaky bark of Shrubby Cinquefoil as tinder for making friction fires.

Trailing Wild Blackberry   *Rubus ursinus*
The Saanich of Vancouver Island used Trailing Wild Blackberry vines to place over and under food in steaming pits, and also for ritual scrubbing. They and other Coast Salish groups sometimes used the fruits as a purple stain.

Sitka Mountain Ash   *Sorbus sitchensis*
The Carrier sometimes made side sticks for snowshoes from Sitka Mountain Ash. One Gitxsan man used the wood to make axe handles. The Dena'ina of Alaska favoured Mountain Ash twigs for steam-bath switches, because the leaves stay on even after drying. They also used the branches of other plants as switches, including Sweet Gale, alders, Trembling Aspen and Highbush Cranberry.

Madder Family   Rubiaceae

Cleavers   *Galium aparine*
The Cowichan rubbed the plants of Cleavers on their hands to remove pitch. They used the dried plants as tinder for lighting fires.

Northern Bedstraw   *Galium boreale*
The Blackfoot of Alberta obtained a red dye from the fine roots of Northern Bedstraw. But if the roots were boiled too long, the colour changed to yellow.

Sweet-scented Bedstraw   *Galium triflorum*
The Blackfoot of Alberta used the dried flowers as a perfume. Ditidaht elder Ida Jones used this plant to make a special hair conditioner that, she said, made your hair grow thick and shiny.

Figwort Family   Scrophulariaceae

Scarlet Paintbrush   *Castilleja miniata*
Unalaska Paintbrush   *Castilleja unalaschensis* x *miniata*
Young Nuxalk girls played a game with paintbrush flowers. They formed two teams and lined up on opposite sides facing each other. One team had a flower and chanted its Nuxalk name. A girl from the other team had to come up and face the singers. If she smiled or laughed, she had to go back, but if she kept a straight face her team got the flower and they sang the song and tried to make a girl from the first team smile. The Kwakwa̱ka'wakw used both species in bouquets.

Unalaska Paintbrush.

Shrubby Penstemon.

Yellow Owl-clover   *Orthocarpus luteus*
The Blackfoot of Alberta used to dye small skins with crushed Yellow Owl-clover. They crushed the plants in full bloom and pressed them firmly on the skin, imparting a reddish-tan colour. They also dyed horsehair and feathers with Yellow Owl-clover.

Wood Betony   *Pedicularis bracteosa*
Nlaka'pamux women used Wood Betony leaves as a pattern for basket designs.

Yellow Penstemon   *Penstemon confertus*
Chelan Penstemon   *Penstemon pruinosus*
The Okanagan boiled penstemon flowers and rubbed them on arrows and other items to give them a blue colouring, said to be indelible.

Shrubby Penstemon   *Penstemon fruticosus*
The Okanagan mashed Shrubby Penstemon leaves and placed them inside moccasins as padding. The Stl'atl'imx rubbed the leafy branches on bunches of Nodding Onions before cooking to "get the whiskers off" and give them a better taste. Secwepemc people used the branches as a flavouring when pit cooking.

Valerian Family   # Valerianaceae

Sitka Valerian   *Valeriana sitchensis*
The Secwepemc used Sitka Valerian as a perfume and disinfectant. They also used to bathe race horses with it. The Gitxsan used a related species, Marsh Valerian, as a perfume for the face and hair.

Sitka Valerian.

APPENDIX 2

Scientific Names of Plants and Animals Mentioned in this Book

Plants

Alpine Larch	*Larix lyallii*
American Bulrush	*Scirpus americanus*
Black Birch	see Water Birch
Blackcap	*Rubus leucodermis*
Black Currant	*Ribes hudsonianum*
Black Spruce	*Picea mariana*
Black Tree Lichen	*Bryoria fremontii*
Bluebunch Wheatgrass	*Agropyron spicatum*
Blue Camas	*Camassia quamash* and *C. leichtlinii*
Blue Clematis	*Clematis columbiana*
Bluejoint	*Calamagrostis canadensis* ssp. *langsdorfii*
Blue Wildrye	*Elymus glaucus*
Boa Kelp	*Egregia menziesii*
brome grasses	*Bromus* species
Brown-eyed Wolf Lichen	*Letharia columbiana*
Brown-stemmed Bog Moss	*Sphagnum lindbergii*
Canada Wildrye	*Elymus canadensis*
Canby's Lovage	*Ligusticum canbyi*
Choke Cherry	*Prunus virginiana*
Cinder Conk	*Inonotus obliquus*
Common Juniper	*Juniperus communis*
Creamy Peavine	*Lathyrus ochroleucus*
Creeping Wildrye	*Elymus triticoides*
Crowberry	*Empetrum nigrum*
"Cut-grass"	see Small-flowered Bulrush
Dunegrass	*Elymus mollis*
Field Wormwood	*Artemisia campestris*

Fowl Mannagrass	*Glyceria striata*
Giant Kelp	*Macrocystis integrifolia*
Inky Cap	*Coprinus micaceus*
Lady Fern	*Athyrium filix-femina*
Large-fruited Desert-parsley	*Lomatium macrocarpum*
Long-flowered Stoneseed	*Lithospermum incisum*
Lowbush Cranberry	*Vaccinium vitis-idaea*
Lyngby's Sedge	*Carex lyngbyei*
Maidenhair Fern	*Adiantum pedatum*
Marsh Horsetail	*Equisetum palustre*
Marsh Valerian	*Valeriana dioica*
Menzies' Tree Moss	*Leucolepis acanthoneuron*
Mountain Birch	see Water Birch
Mountain Gooseberry	*Ribes irriguum*
Mountain Hemlock	*Tsuga mertensiana*
Narrow-leaved Desert-parsley	*Lomatium triternatum*
Needle-and-thread Grass	*Stipa comata*
Nodding Woodreed	*Cinna latifolia*
Old Man's Beard	*Alectoria sarmentosa*
Old Man's Beard	*Usnea longissima*
Oregon Woodsia	*Woodsia oregana*
Pin Cherry	*Prunus pensylvanica*
Poison Hemlock	*Conium maculatum*
Porcupine-grass	*Stipa spartea*
puffballs	*Lycoperdon* species, *Calvatia gigantea* and *Bovista* species
Red Laver	*Porphyra abbottae*
Raphia	*Raphia farinifera*
Rough Moss	*Claopodium crispifolium*
Running Clubmoss	*Lycopodium clavatum*
Sac Seaweed	*Halosaccion glandiforme*
Sea Wrack	*Fucus gardneri*
Sitka Willow	*Salix sitchensis*
Small-flowered Bulrush	*Scirpus microcarpus*
Small Red Peat Moss	*Sphagnum capillifolium*
"Speargrass"	see Needle-and-thread Grass
Spiny Wood Fern	*Dryopteris expansa*
Spread-leaved Peat Moss	*Sphagnum squarrosum*
Swamp Birch	*Betula pumila* var. *glandulifera*
Sweet Gale	*Myrica gale*
"Sweetgrass"	see American Bulrush

"Three Square"	see American Bulrush
Tinderwood Polypore	*Fomes fomentarius*
Trailing Black Currant	*Ribes laxiflorum*
Tufted Hairgrass	*Deschampsia caespitosa*
Turkey Tails	*Coriolus versicolor*
Water Birch	*Betula occidentalis*
Water Parsley	*Oenanthe sarmentosa*
Waxy Currant	*Ribes cereum*
Whitebark Pine	*Pinus albicaulis*
White-stemmed Gooseberry	*Ribes inerme*
Wild Raspberry	*Rubus idaeus*

Animals

Beaver	*Castor canadensis*
Black Bear	*Ursus americanus*
Black Cod	see Sablefish
Butter Clam	*Saxidomus giganteus*
Caribou	*Rangifer tarandus*
Chinook Salmon	*Oncorhynchus tshawytscha*
Elk	*Cervus elaphus*
Eulachon	*Thaleichthys pacificus*
Lingcod	*Ophiodon elongatus*
Moose	*Alces alces*
Mountain Goat	*Oreamnos americanus*
Sablefish	*Anoplopoma fimbria*
Sockeye Salmon	*Oncorhynchus nerka*

GLOSSARY

Algae (singular, alga) A large group of plants, mostly aquatic or marine, that have no true roots, stems, leaves or specialized conduction tissue; includes seaweeds.

Alternate With a single leaf or bud at each node along a stem (as opposed to *opposite*).

Axil The upper angle between a leaf and a stem.

Basal At or emerging from the base of a plant or structure.

Bract A modified leaf, either small and scale-like or large and petal-like.

Bryophyte Any member of the plant division Bryophyta, comprising the mosses and liverworts.

Cambium A layer of continuously dividing cells between the wood and the bark of trees and shrubs, from which new wood and bark tissues are derived.

Catkin A drooping, elongated cluster of minute flowers, either male or female, as on willows, alders and birches.

Chlorophyll The characteristic green pigment of plants, an essential pigment (or light absorber) in photosynthesis.

Compound Composed of two or more similar parts. A compound leaf is divided into two or more leaflets with a common leafstalk.

Cone   A reproductive structure, either male or female, of certain trees, consisting of a central axis surrounded by numerous woody scales that bear the seeds or pollen; e.g., a pine cone.

Conifer   Any cone-bearing tree such as pine, fir or spruce; a major group of gymnosperms.

Cordage   Cords, ropes and strings, including fishing lines and spun fibres prepared for weaving.

Crown   The leafy or branching part of a tree.

Deciduous   Refers to a plant that sheds all its leaves annually.

Dicotyledon   Any member of a major subgroup of flowering plants (Dicotyledonae) characterized by embryos with two seed-leaves (cotyledons), net-veined leaves and flower parts in fours or fives. (See Monocotyledon.)

Diatomaceous earth   A whitish, calcium rich earth composed of ancient deposits of microscopic algae called diatoms.

Evergreen   Refers to a plant that keeps its green leaves throughout the year, even during the winter.

Family   A category in the classification of plants and animals, ranking above a genus and below an order; including two or more related genera. Most family names end in "aceae."

Fern   Any member of a broad division of non-flowering plants (Pteridophyta) that have true roots, stems, specialized conduction tissue and true leaves, which are usually large and compound or dissected. Ferns reproduce by spores, usually produced in sori on the lower surfaces or margins of the leaves.

Flowering Plant   Any member of a major group of vascular plants known as angiosperms (Magnoliophyta), characterized by having true flowers and seeds enclosed in a fruit.

Fertile   Capable of producing viable seed, or as applied to stamens, capable of producing viable pollen.

Fruit A ripened seed case or ovary and any associated structures that ripen with it.

Fungi (singular, fungus) A broad group of organisms, generally considered distinct from plants, lacking chlorophyll and true roots, stems and leaves; includes moulds, mildews, rusts, smuts and mushrooms. Fungi reproduce by spores.

Genus (plural, genera) The main subdivision of family in the classification of plants and animals, consisting of a group of closely related species. In the scientific name of an organism, the genus name is the first term, and the initial letter is always capitalized; e.g., *Pinus* is the genus name in *Pinus contorta* (Lodgepole Pine).

Gymnosperm Any member of a major group of vascular plants (Pinophyta) characterized by having seeds or ovules that are not enclosed in a fruit but borne in cones or related structures. The conifers are an important subgroup of gymnosperms.

Habit The characteristic form or manner of growth of a plant.

Haida Gwaii The Queen Charlotte Islands.

Herbaceous Not woody; having stems that die back to the ground at the end of the growing season.

Holdfast A structure, usually with branching, rootlike appendages, by which a seaweed is fastened to the surface it is growing on.

Imbrication A decoration in basketry made by overlapping strands of fibre over the weave of the basket. A skilled weaver can create intricate designs by imbricating with coloured bark, grasses, etc.

Kerf A groove. Box makers cut kerfs in cedar boards to make bent-wood boxes (see the account for Western Red-cedar).

Labret An ornament inserted into a perforation in the lower lip; also called a lip plug.

Leaflet One of the units of a compound leaf.

Lenticel Slightly raised area, usually elongated, on the surface of the bark of certain trees and shrubs. The cells are more loosely arranged than in the surrounding tissue.

Lichen Any member of a large group of composite organisms, each consisting of one or more algae and a fungus growing in a close relationship. Lichens are generally small, forming branching, leafy or encrusting structures on rock, wood, bark and soil.

Lobe The major division of a leaf extending about half way to the base or centre; oak and maple leaves are lobed.

Monocotyledon Any member of a major subgroup of flowering plants (Monocotyledonae) characterized by embryos with a single seed-leaf (cotyledon), parallel-veined leaves and flower parts in threes. (See Dicotyledon.)

Mycelium The mass of minute tubular structures making up the main body of a fungus. It is typically embedded in a substrate such as wood or soil.

Node The point on a stem where one or more leaves or branches are attached.

Nut A hard, dry fruit that remains closed at maturity.

Opposite Growing directly across from each other at the same node – as opposed to alternate.

Palmately Arising from the same point. A palmately compound leaf has leaflets that grow from the same point; e.g., a lupine leaf.

Perennial A plant that lives more than two years.

Petal Any member of the inside set of floral bracts in flowering plants; usually coloured or white and serving to attract insect or bird pollinators. Many flowers do not have true petals.

Pinnae (singular, Pinna) The primary lateral divisions of a pinnately compound leaf, such as that of a fern.

Pinnately compound Referring to a compound leaf with leaflets on either side of a central axis in a feather-like arrangement; e.g., the leaf of an elderberry.

Pinnule An ultimate leaflet of a leaf that is pinnately compound two or more times; i.e., the ultimate division of a compound pinna.

Pistillate Refers to flowering structures having one or more pistils but no stamens. The pistil is the female or seed-bearing organ of a flower. (See Staminate.)

Pollen The mass of young male reproductive bodies (pollen grains) of a seed plant at the stage when they are released from the anther or pollen capsule.

Pores Minute crowded holes or tubules characteristic of pore fungi; these bear spore-producing structures on their surfaces.

Rhizome A creeping underground stem, often fleshy, serving in vegetative reproduction and food storage.

Sepal Any member of the outside set of floral bracts in flowering plants; typically green and leaflike, but sometimes brightly coloured and petal-like.

Shrub A relatively small woody perennial, usually with several permanent stems instead of a single trunk, like that of a tree.

Silicon A non-metallic element that, combined with oxygen as silicon dioxide, is a major component of such substances as quartz, opal and sand.

Sori (singular, sorus) Clusters of spore cases on the undersurface of a fern frond.

Species (singular or plural) The fundamental unit in the classification of plants and animals, a subdivision of a genus; consisting of a group of organisms that have a high degree of similarity, show persistent differences from members of species in the same genus and usually interbreed only among themselves. In a scientific name, the species is designated by the second part, which is not capitalized; e.g., in *Pinus contorta* (Lodgepole Pine), *contorta* is the species designation.

Spike An elongated flower cluster, with flowers attached directly to the central stalk.

Spikelet Diminutive of spike; an ultimate flowering unit in a compound flower cluster, especially in grasses.

Spore A one-celled reproductive structure in non-flowering plants such as mosses and ferns.

Staminate Refers to flowering structures having one or more stamens but no pistils. The stamen is the male or pollen-bearing organ of a flower. (See Pistillate.)

Stipe An erect stemlike portion of a seaweed.

Terminal Growing at the end of a stem or branch.

Tree A (large) woody perennial having a single main stem or trunk.

Tumpline A woven strap worn across the forehead and used to carry pack-baskets or other containers on the back.

Vegetative Relating to plants or parts of plants lacking reproductive structures.

Warp Strands of fibre stretched lengthwise, to be crossed by the weft, in the making of fabric, mats, baskets, etc.

Weft Transverse strands of fibre woven across a warp to make fabric, matting, baskets, etc.

Whorl A ring of three or more leaves or branches growing from the same point on a stem.

Withe A tough, flexible twig, especially from a willow.

REFERENCES

Anderson, M.K. 1993a. Native Californians as Ancient and Contemporary Cultivators. In *Before the Wilderness. Environmental Management by Native Californians,* edited by T.C. Blackburn and M.K. Anderson. Menlo Park, CA: Ballena Press.

——. 1993b. California Indian Horticulture: Management and Use of Redbud by the Southern Sierra Miwok. *Journal of Ethnobiology* 11:1:145–57.

——. 1993c. The Experimental Approach to Assessment of the Potential Ecological Effects of Horticultural Practices by Indigenous Peoples on California Wildlands. Ph.D. diss., Wildland Resource Science, University of California at Berkeley.

Bandoni, R.J., and A.F. Szczawinski. 1976. *Guide to Common Mushrooms of British Columbia.* Handbook no. 24. Victoria: British Columbia Provincial Museum.

Barnett, H.G. 1955. *The Coast Salish of British Columbia.* Eugene: University of Oregon Press.

Bernick, K. 1991. Wet site archaeology in the Lower Mainland region of British Columbia. Report Prepared for Heritage Trust, Heritage Conservation Branch, Government of British Columbia, Victoria.

——, ed. 1998. *Hidden Dimensions: The Cultural Significance of Wetland Archaeology.* Vancouver: UBC Press.

Blackburn, T.C., and M.K. Anderson, eds. 1994. *Before the Wilderness. Environmental Management by Native Californians.* Menlo Park, CA: Ballena Press.

Blackman, M.B. 1982. *During My Time: Florence Edenshaw Davidson, a Haida Woman.* Seattle: University of Washington Press.

Blanchette, R.A., B.D. Compton, N.J. Turner and R.L. Gilbertson. 1992. Nineteenth century shaman grave guardians are carved *Fomitopsis officinalis* sporophores. *Mycologia* 84:1:119–24.

Boas, F. 1909. The Kwakiutl of Vancouver Island. In *Publications of the Jesup North Pacific Expedition,* part 2.

——. 1921. *Ethnology of the Kwakiutl.* Bureau of American Ethnology, 35th

Annual Report, Pt. 1, 1913–14. Washington, DC: Smithsonian Institution.

Bouchard, R. 1973. Mainland Comox Plant Names. Unpublished manuscript. British Columbia Indian Language Project, Victoria.

Boyd, L. 1990. *For Someone Special*. Developing Our Resources Curriculum Project, sponsored by Quesnel School District's Native Education Program, Quesnel, B.C.

Boyd, R.T. 1990. Demographic history, 1774-1874. In *Northwest Coast*, vol. 7 of *Handbook of North American Indians*, edited by W. Suttles (W.C. Sturtevant, General Editor). Washington, DC: Smithsonian Institution.

Brayshaw, T.C. 1985. *Pondweeds and Bur-reeds, and Their Relatives, of British Columbia*. Occasional Paper no. 26. Victoria: British Columbia Provincial Museum.

——. 1989. *Buttercups, Waterlilies and Their Relatives in British Columbia*. Memoir no. 1. Victoria: Royal British Columbia Museum.

——. 1996a. *Catkin-Bearing Plants of British Columbia*. Victoria: Royal British Columbia Museum.

——. 1996b. *Trees and Shrubs of British Columbia*. Royal British Columbia Museum Handbook. Vancouver: UBC Press; Victoria: Royal British Columbia Museum.

Calder, J.A., and R.L. Taylor.1968. *Flora of the Queen Charlotte Islands*. Part 1. Monograph no. 4. Ottawa: Canada Department of Agriculture, Research Branch.

California Indian Basketweavers Association. 1996. *From the Roots: California Indian Basketweavers*. Videocassette. CIBA, Nevada City, CA.

Carrier Linguistic Committee. 1973. *Hanúyeh Ghun 'Útni-i. (Plants of Carrier Country)*. (Central Carrier Language.) Fort St James, B.C.: Carrier Linguistic Committee.

Chamberlain, A.B. 1892. Report on the Kootenay Indians of Southeastern British Columbia. *Eighth Report on the Northwestern Tribes of Canada*. British Association for the Advancement of Science, Edinburgh Meeting.

Claxton, E., and J. Elliott. 1994. *Reef Net Technology of the Saltwater People*. Brentwood Bay, B.C.: Saanich Indian School Board.

Cole, D., and B. Lockner, eds. 1989. *The Journals of George M. Dawson: British Columbia, 1875–1878*, vols 1 and 2. Vancouver: UBC Press.

Compton, B.D. 1993. Upper North Wakashan and Southern Tsimshian ethnobotany: the knowledge and usage of plants and fungi among the Oweekeno, Hanaksiala (Kitlope and Kemano), Haisla (Kitamaat) and Kitasoo peoples of the central and north coasts of British Columbia. Ph.D. diss. Department of Botany, University of British Columbia.

Coville, F.V.1904. Plants used in basketry. In *Aborginal American Basketry*, edited by O.T. Mason. Report of the U.S. National Museum. Washington, DC: Government Printing Office.

Craig, J., and R. Smith. 1997. *A Rich Forest: Traditional Knowledge, Inventory and*

Restoration of Culturally Important Plants and Habitats in the Atleo River Watershed. Final Report of the Ahousaht Ethnobotany Project, 1996. Ahousaht, B.C.: Ahousaht Band Council; Tofino, B.C.: Long Beach Model Forest.

Croes, D.R., ed. 1976. *The Excavation of Water-Saturated Archaeological Sites (Wet Sites) on the Northwest Coast of North America.* Mercury Series, Archaeological Survey of Canada Paper No. 50. Ottawa: National Museum of Man.

———. 1977. Basketry from the Ozette Village Archaeological Site: a technological, functional and comparative study. Ph.D. diss., Washington State University, Pullman. (Ann Arbor, Michgan: University Microfilms 77-25, 762.)

———. 1989. Prehistoric ethnicity on the Northwest Coast of North America: an evaluation of style in basketry and lithics. *Journal of Anthropological Archaeology* 8:101–30.

———. 1995. *The Hoko River Archaeological Site Complex.* Pullman: Washington State University Press.

Davidson, J. 1927. *Conifers, Junipers and Yew: Gymnosperms of British Columbia.* London: T. Fisher Unwin Ltd.

Dawson, G.M. 1891. Notes on the Shuswap people of British Columbia. *Transactions of the Royal Society of Canada*, Section II, Part I, pp. 3–44.

Douglas, G.W., G.B. Straley and D. Meidinger. 1989–94. *The Vascular Plants of British Columbia*, 4 vols. Victoria: British Columbia Ministry of Forests.

Driver, H.E.1961. *Indians of North America.* Chicago: University of Chicago Press.

Drucker, P. 1951. *The Northern and Central Nootkan Tribes.* Bureau of American Ethnology, Bulletin 44. Washington, DC: Smithsonian Institution.

———. 1955. *Indians of the Northwest Coast.* Garden City, NY: Natural History Press.

———. 1965. *Cultures of the North Pacific Coast.* San Francisco: Chandler Publishing Company.

Duff, W. 1952. *The Upper Stalo Indians.* Anthropology in British Columbia, Memoir no. 1. Victoria: British Columbia Provincial Museum.

———. 1964. *The Indian History of British Columbia: The Impact of the White Man.* Anthropology in British Columbia, Memoir no. 5. Victoria: British Columbia Provincial Museum. New edition, reprinted, 1997. Victoria: Royal British Columbia Museum.

Duff, W., and M. Kew. 1973. A Select Bibliography of Anthropology in British Columbia. Revised by F. Woodward and L. Ruus. *B.C. Studies* 19:73–122. (See also Hoover 1986.)

Ellis, D.W., and L. Swan. 1981. *Teachings of the Tides. Uses of Marine Invertebrates by the Manhousat People.* Nanaimo, B.C.: Theytus Books.

Ellis, D.W., and S. Wilson. 1981. *The Knowledge and Usage of Marine Invertebrates by the Skidegate Haida People of the Queen Charlotte Islands.* Skidegate, B.C.: Queen Charlotte Islands Museum Society.

Emmons, G.T. 1903. *The Basketry of the Tlingit.* Memoirs of the American Museum of Natural History, vol. 3 (2). New York: American Museum of Natural History.

——. 1911. *The Tahltan Indians.* Anthropological Publications, vol. 4 (1). Philadelphia: University of Pennsylvania (The Museum).

Fladmark, K.R. 1986. *British Columbia Prehistory.* Ottawa: National Museum of Man, National Museums of Canada.

Forests, British Columbia Ministry of. 1992. Biogeoclimatic Zones of British Columbia. Map. Victoria: British Columbia Ministry of Forests, Research Branch.

——. 1997. *Culturally Modified Trees of British Columbia.* A Handbook for the Identification and Recording of Culturally Modified Trees. Victoria: British Columbia Ministry of Forests.

Friedman, J. 1975. The prehistoric uses of wood at the Ozette Archeological Site. Ph.D. diss., Department of Anthropology, University of Washington.

——. 1978. *Wood Identification by Microscopic Examination.* Heritage Record no. 5. Victoria: British Columbia Provincial Museum.

Gero, J.M., and M.K. Conkey, eds. 1991. *Engendering Archaeology: Women and Prehistory.* Oxford, UK, and Cambridge, MA: B. Blackwell.

——. 1984. American Indian art: values and aesthetics. *American Indian Basketry Magazine* 4:4:4–30.

Gottesfeld, L.M.J. 1992a. Short communication: use of Cinder Conk (*Inonotus obliquus*) by the Gitksan of Northwestern British Columbia, Canada. *Journal of Ethnobiology* 12:1:153–56.

——. 1992b. The importance of bark products in the aboriginal economies of northwestern British Columbia, Canada. *Economic Botany* 46:2:148–57.

——. 1994a. Aboriginal burning for vegetation management in northwest British Columbia. *Human Ecology* 22:2:171–88.

——. 1994b. Conservation, territory and traditional beliefs: an analysis of Gitksan and Wet'suwet'en subsistence, northwest British Columbia, Canada. *Human Ecology* 22:4:443–65.

——. 1994c. Wet'suwet'en ethnobotany. *Journal of Ethnobiology* 14:185–210.

Gottesfeld, L.M.J., and D.H. Vitt. 1996. The selection of sphagnum for diapers by indigenous North Americans. *Evansia* 13:3:103–8.

Gunther, E. 1945. *Ethnobotany of Western Washington.* University of Washington Publications in Anthropology, vol. 10, no. 1. Reprinted 1973. Seattle: University of Washington Press.

Gustafson, Paul. 1980. *Salish Weaving.* Vancouver: Douglas & McIntyre; Seattle: University of Washington Press.

Haeberlin, H.K., J.A. Teit and H.H. Roberts. 1928. Coiled basketry in British Columbia and surrounding region. *Bureau of American Ethnology, 41st Annual Report,* pp. 119–484. Washington, DC: Smithsonian Institution.

Hart, J.A. 1974. Plant taxonomy of the Salish and Kootenai Indians of western Montana. M.A. thesis. University of Montana, Missoula.

Hart, J.A. 1979. The ethnobotany of the Flathead Indians of western Montana. *Botanical Museum Leaflets* (Harvard University) 27:10:261–307.

Hart, J.A., N.J. Turner and L. Morgan. 1981. Ethnobotany of the Kootenai [Ktunaxa] Indians of western North America. Report to the Kootenay Indian Area Council (now the Ktunaxa Kinbasket Tribal Council), Cranbrook, B.C.

Hartzell, H. 1991. *The Yew Tree. A Thousand Whispers.* Eugene, OR: Hulogosi.

Hayden, B., ed. 1992. *Complex Cultures of the British Columbia Plateau: Traditional Stl'atl'imx Resource Use.* Vancouver: UBC Press.

Hebda, R.J., and R.W. Mathewes. 1984. Holocene history of cedar and native Indian cultures of the North American Pacific coast. *Science* 225:711–13.

Hebda, R.J., N.J. Turner, S. Birchwater, M. Kay and the Elders of Ulkatcho. 1996. *Ulkatcho Food and Medicine Plants.* Anahim Lake, B.C.: Ulkatcho Indian Band.

Hellson, J.C., and M. Gadd.1974. *Ethnobotany of the Blackfoot Indians.* National Museum of Man. Mercury Series. Canadian Ethnology Service Paper no. 19. Ottawa: National Museum of Man, National Museums of Canada.

Helm, J., ed. 1981. *Subarctic.* Vol. 6. of *Handbook of North American Indians* (W.C. Sturtevant, General Editor). Washington, DC.: Smithsonian Institution.

Hitchcock, C. L., A. Cronquist, M. Ownbey and J.W. Thompson. 1955–69. *Vascular Plants of the Pacific Northwest,* 5 vols. Seattle: University of Washington Press.

Hoover, A.L. 1986. A selection of publications on the Indians of British Columbia. Unpublished Bibliography. Royal British Columbia Museum Library, Victoria. (An Update of Duff and Kew, 1973).

Hubbard, W.A. 1969. *The Grasses of British Columbia.* Handbook no. 9. Victoria: British Columbia Provincial Museum.

Hultén, E. 1968. *Flora of Alaska and Neighboring Territories.* Stanford, CA: Stanford University Press.

Hunn, E.S., with James Selam and family. 1990. *Nch'i-Wana. "The Big River."* *Mid-Columbia Indians and Their Land.* Seattle: University of Washington Press.

Hunn, E.S., N.J. Turner and D.H. French. 1998 (in press). Ethnobiology and subsistence. In *Plateau,* edited by D.E. Walker. Vol. 12 of *Handbook of North American Indians* (William C. Sturtevant, General Editor). Washington, DC.: Smithsonian Institution.

Jenness, D. (n.d.) The Saanich Indians and Coast Salish field notes. Unpublished manuscripts. (Ethnobotanical excerpts by David Rozen, Victoria. National Museum of Man, Ethnology Archives, ms. #1103.6.) National Museums of Canada, Ottawa.

Johnson, D., L. Kershaw, A. MacKinnon and J. Pojar, eds. 1995. *Plants of the Western Boreal Forest and Aspen Parkland.* Vancouver and Edmonton: Lone Pine Publishing.

Johnston, A. 1970. Blackfoot Indian utilization of the flora of the northwestern Great Plains. *Economic Botany* 24:3:301–24.

Kari, P.R. 1987. *Tanaina Plantlore. Dena'ina K'et'una. An Ethnobotany of the Dena'ina Indians of South-central Alaska.* Anchorage, AK: National Park Service, Alaska Region.

Kay, M.S. 1995. Environmental, cultural and linguistic factors affecting Ulkatcho (Carrier) botanical knowledge. M.Sc. thesis, Department of Biology, University of Victoria.

Kennedy, D.I.D., and R. Bouchard. 1974. Utilization of fishes, beach foods and marine animals by the Tl'úluus Indian people of British Columbia. Unpublished manuscript. British Columbia Indian Language Project, Victoria.

——. 1975a. Utilization of fish by the Chase Shuswap Indian people of British Columbia. Unpublished manuscript. British Columbia Indian Language Project, Victoria.

——. 1975b. Utilization of fish by the Colville Okanagan Indian people. Unpublished manuscript. British Columbia Indian Language Project, Victoria.

——. 1975c. Utilization of fish by the Mount Currie Lillooet Indian people of British Columbia. Unpublished manuscript. British Columbia Indian Language Project, Victoria.

——. 1976. Utilization of fish, beach foods, and marine mammals by the Squamish Indian people of British Columbia. Unpublished manuscript. British Columbia Indian Language Project, Victoria.

——. 1983. *Sliammon Life, Sliammon Lands.* Vancouver: Talonbooks.

Kershaw, L., A. MacKinnon and J. Pojar. 1998. *Plants of the Rocky Mountains.* Vancouver and Edmonton: Lone Pine Publishing.

Kirk, R. 1979. *Ozette and Hoko River Archaeology.* Pullman, WA: Archaeological Research Center, Washington State University.

——. 1986. *Wisdom of the Elders: Native Traditions on the Northwest Coast.* Vancouver: Douglas and McIntyre.

Krajina, V.J.1969. Ecology of Forest Trees in British Columbia. *Ecology of Western North America* 2:1.

Kruckeberg, A.R. 1996. *Gardening With Native Plants of the Pacific Northwest.* 2nd ed. Vancouver: Douglas & McIntyre; Seattle: University of Washington Press.

Kuneki, N.J., E. Thomas and M. Lockish. 1982. *The Heritage of Klickitat Basketry.* Portland: Oregon Historical Society Press.

Laforet, A. 1984. Tsimshian basketry. In The Tsimshian. *Images of the Past: Views for the Present,* edited by M. Seguin. Vancouver: UBC Press.

——. 1990. Regional and personal style in Northwest Coast basketry. In *The Art of Native American Basketry: A Living Legacy,* edited by F.W. Porter. Contributions to the Study of Anthropology, no. 5. New York: Greenwood Press.

——. 1992. Windows on diversity: Northwest coast baskets in the Pitt Rivers collection. In Basketmakers. Meaning and Form in Native American Baskets, edited by L. Mowat, H. Murphy and P. Dransart. Monograph 5. Oxford, UK: Pitt Rivers Museum (University of Oxford).

Leechman, D. 1932. Aboriginal paints and dyes in Canada. In *Transactions of the Royal Society of Canada*, Section 2 (1932). Ottawa: The Royal Society of Canada.

Lepofsky, D. In press, a. The northwest. In *Plants and People in Ancient North America*, edited by Paul Minnis. Washington, DC: Smithsonian Institution Press.

———. In press, b. Plants and pithouses: the archaeobotany of complex hunter-gatherers on the British Columbian plateau. In *The Archaeobotany of Hunter-Gatherers*, edited by J. Hather and S. Mason. London, UK: University College Institute of Archaeology.

Lyons, C.P., and Bill Merilees. 1995. *Trees, Shrubs and Flowers to Know in British Columbia and Washington*. Edmonton and Vancouver: Lone Pine Publishing.

MacKinnon, A., J. Pojar and R. Coupé, editors. 1992. *Plants of Northern British Columbia*. Edmonton and Vancouver: Lone Pine Publishing.

McIlwraith, T.F. 1948. *The Bella Coola Indians*. 2 vols. Toronto: University of Toronto Press.

McMillan, A.D. 1988. *Native Peoples and Cultures of Canada: An Anthropological Overview*. Vancouver: Douglas & McIntyre.

McNeary, S. 1974. The traditional economic and social life of the Niska. Unpublished report. Ottawa: National Museum of Man.

———. 1976. Where fire came down: social and economic life of the Niska. Ph.D. thesis. Bryn Mawr College, PA.

Media Resource Associates. 1994. *Indian America. A Gift from the Past. (A film about the Ozette Project and the Makah people of Neah Bay)*. (Narrated by Wes Studi.) Film. Washington, DC: Media Resource Associates.

Meidinger, D., and J. Pojar, eds. 1991. *Ecosystems of British Columbia*. Special Report Series 6. Victoria: British Columbia Ministry of Forests, Research Branch.

Meilleur, B.A., E.S. Hunn and R.L. Cox. 1990. *Lomatium dissectum* (Apiaceae): Multi-purpose plant of the Pacific Northwest. *Journal of Ethnobiology* 10:1:1–20.

Morice, A.G. 1895. Notes archaeological, industrial and sociological on the Western Dénés, with an ethnographical sketch of the same. *Transactions of the Canadian Institute* 4:1–222.

Muckle, R.J. 1998. *The First Nations of British Columbia*. Vancouver: UBC Press.

Nelson, J. 1983. *The Weavers. A Queen Charlotte Islands Reader*. Vancouver: Pacific Educational Press.

Newcombe, C.F. 1902–1910. Unpublished field notes on Nootka, Haida and Salishan groups. Victoria: British Columbia Archives.

Nicholas, G.P. 1994. Prehistoric human ecology as cultural resource management. In *Cultural Resource Management. Archaeological Research, Preservation Planning and Publication Education in the Northeastern United States*, edited by J.E. Kerber. Westport, CT: Bergin & Garvey.

——. 1997. Working together – Archaeology, education and the Secwepemc. *Society for American Archaeology Bulletin* 15:2:8–11.

——. 1998. Wetlands and hunter-gatherer land use in North America. In *Hidden Dimenstions: The Cultural Significance of Wetland Archaeology*, edited by K. Bernick. Vancouver: UBC Press.

Nicholas, G.P., and T.D. Andrews. 1997. *At a Crossroads: Archaeology and First Peoples in Canada*. Burnaby, B.C.: Simon Fraser University, Archaeology Press.

Norton, H.H. 1981. Plant use in Kaigani Haida culture: correction of an ethnohistorical oversight. *Economic Botany* 35:434–49.

Ortiz, B. 1993a. Contemporary California Indian basket-weavers and the environment. In *Before the Wilderness. Environmental Management by Native Californians*, edited by T.C. Blackburn and M.K. Anderson. Menlo Park, CA: Ballena Press.

——. 1993b. Pesticides and basketry. News from Native California. *Skills and Technology*, summer 1993: 7–10.

Palmer, G. 1975. Shuswap Indian Ethnobotany. *Syesis.* 8:29–81.

Parish, R., R. Coupé and D. Lloyd, eds. 1996. *Plants of the Southern Interior, British Columbia.* Edmonton and Vancouver: Lone Pine Publishing.

Paul, F., 1944. *Spruce Root Basketry of the Alaska Tlingit.* Reprinted 1991. Sitka, AK: Sheldon Jackson Museum.

Peacock, S.L. 1992. Piikáni ethnobotany: traditional plant knowledge of the Piikáni peoples of the northwestern plains. M.A. thesis. Department of Archaeology, University of Calgary, Alberta.

Peacock, S.L., and N.J. Turner. In press. Just like a garden: traditional plant resource management and biodiversity conservation on the British Columbia Plateau. In *Biodiversity and Native North America*, edited by P. Minnis and W. Elisens. Norman, OK: University of Oklahoma Press.

Peri, D.W., and S.M. Patterson. 1993. "The basket is in the roots, that's where it begins." *Before the Wilderness. Environmental Management by Native Californians*, edited by T.C. Blackburn and M.K. Anderson. Menlo Park, CA: Ballena Press.

Pojar, J., and A. MacKinnon, eds. 1994. *Plants of Coastal British Columbia, including Washington, Oregon and Alaska.* Vancouver and Edmonton: Lone Pine Publishing.

Porter, F.W., ed. 1990. *The Art of Native American Basketry. A Living Legacy.* Contributions to the Study of Anthropology, no. 5. New York: Greenwood Press.

Samuel, C. 1987. *The Raven's Tail.* Vancouver: UBC Press.

Scagel, R.F. 1967. *Guide to Common Seaweeds of British Columbia.* Handbook no. 27. Victoria: British Columbia Provincial Museum.

Schlick, M.D. 1994. *Columbia River Basketry. Gift of the Ancestors, Gift of the*

Earth. Seattle: University of Washington Press.

Schofield, W.B. 1992. *Some Common Mosses of British Columbia.* 2nd ed. Royal British Columbia Museum Handbook. Victoria: Royal British Columbia Museum.

Scientific Panel for Sustainable Forest Practices in Clayoquot Sound. 1995. *First Nations' Perspectives on Forest Practices in Clayoquot Sound.* Report 3. Victoria: Scientific Panel for Sustainable Forest Practices in Clayoquot Sound.

Secwpemec Cultural Education Society. 1986a. *Shuswap Homes.* Shuswap Cultural Series, Book 3. Kamloops, B.C.: Secwpemec Cultural Education Society.

———. 1986b. *Traditional Shuswap Clothing and Adornment.* Shuswap Cultural Series, Book 4. Kamloops, B.C.: Secwpemec Cultural Education Society.

———. 1986c. *Technology of the Shuswap.* Shuswap Cultural Series, Book 5. Kamloops, B.C.: Secwpemec Cultural Education Society.

Sewid-Smith, Daisy (Mayanilh), and Chief Adam Dick (Kwaxsistala). In press. The sacred cedar tree of the Kwakwaka'wakw people. Interview with Nancy J. Turner. In *Stars Above, Earth Below: Native Americans and Nature,* edited by Marsha Bol. Background book for the *Alocoa Foundation Hall of Native Americans* exhibit. Pittsburgh, PA: The Carnegie Museum of Natural History.

Shackelford, R.S. 1900. Legend of the Klickitat Basket. *American Anthropologist* 2:779–80.

Smith, H.I. 1927. *Handbook of the Kitwanga Garden of Native Plants.* Ottawa: National Museum of Canada and Department of Indian Affairs.

———. 1997. *Ethnobotany of the Gitksan Indians of British Columbia.* Edited, annotated and expanded by B.D. Compton, B. Rigsby and M.L. Tarpent. Mercury Series, Canadian Ethnology Service, Paper 132. Ottawa: Canadian Museum of Civilization.

Sproat, G.M. 1868. *Scenes and Studies of Savage Life.* London: n.p.

Steedman, E.V., ed. 1930. The Ethnobotany of the Thompson Indians of British Columbia. (Based on James Teit's field notes.) In *Bureau of American Ethnology, 45th Annual Report,* 1927–28. Washington, DC: Smithsonian Institution.

Stewart, H. 1984. *Cedar: Tree of Life to the Northwest Coast Indians.* Vancouver: Douglas & McIntyre; Seattle: University of Washington Press.

———. 1977. *Indian Fishing. Early Methods on the Northwest Coast.* Seattle: University of Washington Press.

Suttles, W., ed. 1990. *Northwest Coast.* Vol. 7 of *Handbook of North American Indians* (W.C. Sturtevant, General Editor). Washington, DC: Smithsonian Institution.

Swanton, J. 1905. *Jesup North Pacific Expedition,* vol. 5, part 1. Contributions to the Ethnology of the Haida, Memoir no. 8, part 1. New York: American Museum of Natural History.

Szczawinski, A.F. 1962. *The Heather Family (Ericaceae) of British Columbia.* Handbook no. 19. Victoria: British Columbia Provincial Museum.

Taylor, T.M.C. 1956. *The Ferns and Fern-allies of British Columbia.* Handbook no. 12. Victoria: British Columbia Provincial Museum.

——. 1966. *The Lily Family (Liliaceae) of British Columbia.* Handbook no. 25. Victoria: British Columbia Provincial Museum.

——. 1973. *The Rose Family (Rosaceae) of British Columbia.* Handbook no. 30. Victoria: British Columbia Provincial Museum.

——. 1974a. *The Pea Family (Leguminosae) of British Columbia.* Handbook no. 32. Victoria: British Columbia Provincial Museum.

——. 1974b. *The Figwort Family (Scrophulariaceae) of British Columbia.* Handbook no. 33. Victoria: British Columbia Provincial Museum.

Teit. J.A. 1900. *The Thompson Indians.* Memoir no. 2. New York: American Museum of Natural History.

——. 1906a. *The Lillooet Indians.* Memoir no. 4. New York: American Museum of Natural History.

——. 1906b. Notes on the Tahltan Indians of British Columbia. In *Boas Anniversary Volume.* New York: American Musuem of Natural History.

——. 1909. *The Shuswap.* Memoir no. 5. New York: American Musuem of Natural History.

——. 1930. The Salishan tribes of the western plateaus. In *Bureau of American Ethnology, 45th Annual Report, 1927–28.* Washington, DC: Smithsonian Institution.

Tepper, L.H., ed. 1987. *The Interior Salish Tribes of British Columbia: A Photographic Collection.* Mercury Series, paper no. 111. Ottawa: Canadian Museum of Civilization, Canadian Ethnology Service.

——, ed. 1991. *The Bella Coola Valley: Harlan I. Smith's Fieldwork Photographs, 1920–1924.* Mercury Series, paper no. 123. Ottawa: Canadian Museum of Civilization, Canadian Ethnology Service.

Thomas, M. 1996. *Birch Bark Baskets.* Salmon Arm, B.C.: Mary Thomas and Family, Neskonlith Band.

Turner, N.J. 1973. The ethnobotany of the Bella Coola Indians of British Columbia. *Syesis* 6:193–220.

——. 1974. Plant taxonomic systems and ethnobotany of three contemporary Indian groups of the Pacific Northwest (Haida, Bella Coola and Lillooet). *Syesis* 7, supplement 1.

——. 1977. Economic importance of Black Tree Lichen (*Bryoria fremontii*) to the Indians of western North America. *Economic Botany* 31:461–70.

——. 1981. Indian use of *Shepherdia canadensis* (Soapberry) in western North America. *Davidsonia* 12:1:1–14.

——. 1982 Traditional use of Devil's Club (*Oplopanax horridus*) by native peoples in western North America. *Journal of Ethnobiology* 2:1:17–38.

——. 1987. General plant categories in Thompson and Lillooet, two Interior Salish languages of British Columbia. *Journal of Ethnobiology* 7:1:55–82.

——. 1988a. Ethnobotany of coniferous trees in Thompson and Lillooet

Interior Salish of British Columbia. *Economic Botany* 42:2:177–94.

——. 1988b. "The importance of a rose." Evaluating the cultural significance of plants in Thompson and Lillooet Interior Salish. *American Anthropologist* 90:2:272–90.

——. 1989. "All berries have relations": midlevel folk plant categories in Thompson and Lillooet Interior Salish. *Journal of Ethnobiology* 9:1:69–110.

——. 1991a. Wild berries. In *Berries*, edited by J. Bennett. Toronto: Harrowsmith Books.

——. 1991b. Burning mountain sides for better crops: aboriginal landscape burning in British Columbia. *Archaeology in Montana* 32:2 (special issue). Excerpts reprinted in *International Journal of Ecoforestry* 10:3:116–22.

——. 1992a. Plant resources of the Stl'atl'imx (Fraser River Lillooet) people: a window into the past. In *Complex Cultures of the British Columbia Plateau: Traditional Stl'atl'imx Resource Use*, edited by B. Hayden. Vancouver: UBC Press.

——. 1992b. "Just when the wild roses bloom": the legacy of a Lillooet basket weaver. *TEK TALK* (A Newsletter of Traditional Ecological Knowledge, UNESCO, World Congress for Education & Communication on Environment & Development) 1:2:5–7.

——, ed. 1992c. *Plants for All Reasons: Culturally Important Plants of Aboriginal Peoples of Southern Vancouver Island*. Teaching manual written by students of Environmental Studies 400C class and University Extension class, July 1992. Victoria: Environmental Studies Program, University of Victoria.

——. 1995a. *Food Plants of Coastal First Peoples*. Royal British Columbia Museum Handbook. Vancouver: UBC Press.

——. 1995b. Ethnobotany today in northwestern North America. In *Ethnobotany: Evolution of a Discipline*, edited by R. Evans Schultes and S. Von Reis. Portland, OR: Dioscorides Press.

——. 1995c. *Some Common Plants of Haida Gwaii: xàadlaa gwaayee guud gina q'aws*. A Handbook for the Haida Gwaii Watchman Program for Gwaii Haanas National Park Reserve & Haida Heritage Site, Haida Gwaii.

——. 1996a. Traditional ecological knowledge. In *The Rain Forests of Home. Profile of a North American Bioregion*, edited by P.K. Schoonmaker, B. von Hagen and E.C. Wolf, Ecotrust. Covelo, CA, and Washington, DC: Island Press.

——. 1996b. *"Dans une Hotte"*. *L'importance de la vannerie das l'économie des peuples chasseurs-pêcheurs-cueilleurs du Nord-Ouest de l'Amérique du Nord*; ("Into a Basket Carried on the Back": Importance of Basketry in Foraging/Hunting/Fishing Economies in Northwestern North America.) *Anthropologie et Sociétiés* 20:3:55–84. (Special issue on Contemporary Ecological Anthropology. Theories, Methods and Research Fields.)

——. 1997a. Traditional Ecological Knowledge. In *The Rain Forests of Home: Profile of a North American Bioregion*, edited by P.K. Schoonmaker, B. Von

Hagen and E.C. Wolf, Ecotrust. Covelo, CA, and Washington, DC: Island Press.

———. 1997b. *"Le fruit de l'ours": Les rapports entre les plantes et les animaux das les langues et les cultures amérindiennes de la Côte-Ouest.* ("The bear's own berry": ethnobotanical knowledge as a reflection of plant/animal interrelationships in northwestern North America). *Recherches amérindiennes au Québec* 27:3/4:31–48 (special edition on *Des Plantes et des Animaux: Visions et Pratiques Autochtones,* edited by P. Beaucage.)

———. 1997c. *Food Plants of Interior First Peoples.* Royal British Columbia Museum Handbook. Vancouver: UBC Press.

———, ed. 1998. *"Making it With Your Hands": Projects Using Indigenous Plant Materials from British Columbia* (by the students of Environmental Studies 416, Fall, 1997). Victoria: School of Environmental Studies, University of Victoria.

Turner, N.J., and M.A.M. Bell. 1971. The Ethnobotany of the Coast Salish Indians of Vancouver Island. *Economic Botany* 25:1:63–104.

———. 1973. The Ethnobotany of the Southern Kwakiutl Indians of British Columbia. *Economic Botany* 27:3:257–310.

Turner, N.J., R. Bouchard and D.I.D. Kennedy. 1980. *The Ethnobotany of the Okanagan-Colville Indians of British Columbia and Washington.* Victoria: British Columbia Provincial Museum.

Turner, N.J., and A. Davis. 1993. "When everything was scarce." The role of plants as famine foods in northwestern North America. *Journal of Ethnobiology* 13:2:1–28.

Turner, N.J., and B.S. Efrat. 1982. *Ethnobotany of the Hesquiat Indians of Vancouver Island.* Cultural Recovery Paper no. 2; Ethnobotanical Contribution no. 1 of the Hesquiat Cultural Committee. Victoria: British Columbia Provincial Museum.

Turner, N.J., and M. Ignace. 1990–97. Secwepemc ethnobotany field notes and manuscripts. Kamloops, B.C.: Secwepemc Cultural Education Society.

Turner, N.J., and D.C. Loewen. In press. The original "free trade": exchange of botanical products and associated plant knowledge in northwestern North America. *Culture* (special edition), edited by Daniel Clément.

Turner, N.J., and A.F. Szczawinski. 1991. *Common Poisonous Plants and Mushrooms of North America.* Portland, OR: Timber Press.

Turner, N.J., J. Thomas, B.F. Carlson and R.T. Ogilvie. 1983. *Ethnobotany of the Nitinaht Indians of Vancouver Island.* Occasional Paper no. 24. Victoria: British Columbia Provincial Museum.

Turner, N.J., L.C. Thompson, M.T. Thompson and A.Z. York. 1990. *Thompson Ethnobotany: Knowledge and Usage of Plants by the Thompson Indians of British Columbia.* Victoria: Royal British Columbia Museum.

U'mista Cultural Society, J. Pasco and B.D. Compton. 1998. *The Living World. Plants and Animals of the Kwakwaka'wakw.* (Kwak'wala edited by Lorraine Hunt.) Alert Bay, B.C.: U'mista Cultural Society.

Vance, N., and J. Thomas, eds. 1997. *Special Forest Products. Biodiversity Meets the Marketplace.* Oregon State University, Sustainable Forestry Seminar Series. Washington, DC: U.S. Department of Agriculture.

Walker , D.E., ed. 1998. *Plateau.* Vol. 12 of *Handbook of North American Indians* (W.C. Sturtevant, General Editor). Washington, DC: Smithsonian Institution.

INDEX

Usage

drying racks, sticks, strings 64, 67, 71, 72, 75, 85, 95, 96, 97, 104, 109, 113, 118, 120, 130, 143, 165, 168, 179, 182, 190, 192, 193, 197, 200, 202, 208

dyes/pigments 46, 49-50, 51, 52, 53, 57, 60, 62, 75, 81, 88, 91, 97, 98, 107, 110, 111-12, 113, 140, 148-49, 150-51, 152, 156, 158, 159, 160, 162, 167, 184, 187, 196, 202, 204, 206, 207, 210, 211, 212, 213, 217, 218, 219, 221; *see also* paint/stains

firewood – *see* fuel

fishing tools 47, 67, 69, 71, 74, 75, 85, 88, 96, 97, 98, 116, 119, 127, 131, 138, 140, 150, 152, 153, 156, 157, 167-68, 169, 172, 174, 175, 179, 182, 183, 185, 189, 195, 196, 200, 201, 202, 204

fishing hooks 87, 88, 91, 96, 97, 98, 101, 102, 140, 157, 160, 171, 173, 180

fuel (wood) 55, 70, 72, 77, 80, 82, 87, 91, 94, 96, 112, 128, 129, 131, 145, 150, 152, 156, 168, 185-86, 186, 195, 198, 200, 214, 217

fumigants – *see* incense

games/toys 45, 48, 54, 62, 64, 69, 85, 96, 97, 101, 104, 109, 112, 115, 125, 130, 132, 146, 164, 180, 186, 191, 200, 214, 218

glue 85, 87, 91, 99, 185, 196

handles, wooden (for tools) 67, 71, 87, 100, 101, 104, 128, 129, 131, 150, 152, 156, 165, 166, 177, 178, 180-81, 182, 183, 185, 186, 189-90, 218, 219

harpoons 71, 88, 91, 96, 97, 100, 129, 166, 178, 182, 185, 191

hats 67, 68, 75, 77, 78, 89, 107, 112, 119, 122, 125, 156, 196

imbrication – *see* basket design

incense/fumigants 49, 69, 83, 117, 118, 135, 143, 144, 159, 186, 218

insecticides 52, 132, 134, 147, 217

insect repellent 53, 55, 83, 113, 117, 134, 135, 142, 143, 144, 147, 206, 206-07, 208, 209, 214, 215, 218

masks 49, 67, 68, 71, 77, 78, 129, 150, 152, 156, 195

mats/matting 48, 68, 75, 77-78, 96, 108-9, 109-10, 114, 116, 119, 120, 122-23, 130, 138, 143, 144, 153, 159, 161, 168, 169, 177, 179, 185, 201, 202, 212

paddles 67, 71, 72, 78, 100, 129, 131, 152, 171, 182

paint/stains 50, 51, 53, 60, 84, 89, 98-99, 102, 112, 123, 128, 140, 149, 151-52, 159, 160, 162, 168, 179, 181, 185, 186, 190, 191, 196; *see also* dyes/pigments

perfumes 68, 117, 118, 189, 196, 209, 215, 217, 218, 220, 221; *see also* deodorant *and* sachets

poison 134, 158-59, 207, 208, 217

sachets 70, 81, 82, 135, 136, 152, 206, 208, 210, 211, 212, 213, 214, 216, 219; *see also* perfumes

scent removal – *see* deodorant

scents – *see* perfumes

shoes 49, 113, 138, 144, 146

smoking food 115, 131, 152, 156, 168, 193, 195, 200, 208

smoking hides 55, 70, 72, 82, 91, 94, 96, 141, 145, 195, 200, 208

snowshoes 69, 82, 85, 94, 96, 101, 128, 129, 131, 152, 156, 174, 199, 200, 201, 210, 219

soap/shampoo/etc. 69, 84, 98, 132, 156, 161, 162, 174, 177, 190, 195-96, 212, 220

Soapberry beaters 113, 114, 130, 131, 168, 170, 212

spears 69, 88, 96, 98, 101, 116, 129, 174, 185

spindle whorls 131

spoons 62, 69, 77, 96, 98, 101, 128, 129, 131, 152, 156, 157, 171, 183, 200

tinder, "slow match" 55, 60, 65, 68, 75, 77, 94, 113, 138, 145, 156, 200, 219

totem poles 67, 72

trapping (terrestrial animals) 74, 77, 138, 139, 201, 204

vessels 75, 76, 82, 97, 151, 152

waterproofing/caulking 56, 88, 91, 93, 96, 97, 100, 176, 185, 196

wedges 71, 87, 98, 100-101, 183

whistles/animal calls 63, 71, 132, 155, 164, 185, 197, 202, 207

This edition was edited, designed and typeset by Gerry Truscott, RBCM. Set in Baskerville (body text: 10/12).

Inside photographic production by Andrew Niemann, RBCM.
Proof reading by Terri Elderton and Tara Steigenberger.
Cover design by Chris Tyrrell, RBCM.

Printed in Canada by Friesens, Winnipeg, Manitoba.